T0340173

Reliability and Maintainability of In-Service Pipelines

Reliability and Maintainability of In-Service Pipelines

Dr. Mojtaba Mahmoodian

Lecturer, RMIT University, Melbourne, VIC, Australia

Gulf Professional Publishing

An imprint of Elsevier

Gulf Professional Publishing is an imprint of Elsevier
50 Hampshire Street, 5th Floor, Cambridge, MA 02139, United States
The Boulevard, Langford Lane, Kidlington, Oxford, OX5 1GB, United Kingdom

Notices
Knowledge and best practice in this field are constantly changing. As new research and experience broaden
our understanding, changes in research methods, professional practices, or medical treatment may become
necessary.

Practitioners and researchers must always rely on their own experience and knowledge in evaluating and
using any information, methods, compounds, or experiments described herein. In using such information
or methods they should be mindful of their own safety and the safety of others, including parties for whom
they have a professional responsibility.

To the fullest extent of the law, neither the Publisher nor the authors, contributors, or editors, assume any
liability for any injury and/or damage to persons or property as a matter of products liability, negligence or
otherwise, or from any use or operation of any methods, products, instructions, or ideas contained in the
material herein.

British Library Cataloguing-in-Publication Data
A catalogue record for this book is available from the British Library

Library of Congress Cataloging-in-Publication Data
A catalog record for this book is available from the Library of Congress

ISBN: 978-0-12-813578-5

For Information on all Gulf Professional Publishing publications
visit our website at https://www.elsevier.com/books-and-journals

**Working together
to grow libraries in
developing countries**

www.elsevier.com • www.bookaid.org

Publisher: Brian Romer
Senior Acquisition Editor: Katie Hammon
Editorial Project Manager: Charlotte Rowley
Production Project Manager: Anitha Sivaraj
Cover Designer: Matthew Limbert

Typeset by MPS Limited, Chennai, India

Dedication

To Fatemeh

Contents

Acknowledgements

Pipelines are valuable assets serving human beings for hundreds of years. Many industries such as oil and gas, and water authorities are dealing with maintenance and management of this infrastructure. Old and deteriorating pipelines frequently fail and cause substantial economic loss, environmental impacts and social damage resulting tremendous cost, inconvenience, and loss of public goodwill.

This publication tries to present trustworthy methodologies for accurate calculation of structural reliability of in-service pipelines and suggest maintenance and inspection approaches.

The author would like to express his deepest appreciation to Prof. Chun Qing Li and Prof. Amir Alani who provided guidance through his academic research on pipeline reliability analysis, since 2010.

The author also acknowledges the inputs and contributions by Vahid Aryai, Aslihan Goru, and Anna Romanova in preparing some sections of the book.

In preparation of this book, some source materials have been used from various standards, handbooks, and technical papers. Acknowledgment is given throughout the book where materials are used.

Dr. Mojtaba Mahmoodian
April 2018

Introduction

Abstract

This chapter introduces the most common type of pipelines used in industry. Pipes categorized based on their material type and their areas of usage as well as their overall characteristics are mentioned.

At the first step of reliability analysis and service life prediction of pipelines, it is necessary to gain knowledge about the design principles and loads affecting buried pipes. The performance criteria for in-service loads are introduced in this chapter as ultimate limit state and serviceability limit state. Flexural and shear failures are the two main ultimate limit states that are considered in design and assessment. Serviceability limit states may be measured by cracking condition. These criteria should be met for a pipeline to be safe and in-service. This chapter reviews the loads and stresses acting on pipelines. The stress formulations will be used later on in reliability analysis, where the limit state function (failure mode) is checked.

Material deterioration is the most common form of pipeline deterioration and is a matter of concern for both the strength and durability functions. How to incorporate the effect of corrosion in the structural analysis of a pipeline is of practical importance. This chapter outlines the corrosion formulation for different pipe types. This knowledge will be used in Chapter 2 for suggesting efficient methods for inspection and maintenance of in-service pipelines as well as in Chapter 4 for formulation of failure function(s) in time-dependent reliability analysis methods.

Reliability and Maintainability of In-Service Pipelines. DOI: https://doi.org/10.1016/B978-0-12-813578-5.00001-9

Chapter Outline

1.1 Background

Pipelines are widely used engineering structures for the collection, conveyance, and distribution of fluid in different areas from rural and urban regions to marine areas. Most of the time, pipelines are placed underground, surcharged by soil weight and traffic loads. Evidently, underground pipelines are required to resist the influence of the external loads (soil and traffic), internal fluid pressure, as well as environmental loads. Buried pipelines are subject to chemical and mechanical loading in their environment of service and these stresses cause failure that is costly to repair.

In many cases underground pipelines are required to withstand particular environmental hazards. Corrosion of pipe material is the most common form of pipeline deterioration and should be considered in both strength and serviceability analysis of buried pipes (Ahammed and Melchers, 1997; Sharma et al., 2008).

According to "The World Factbook" (2010), the United States has approximately 800,000 km and Russia has 252,000 km of pipes transporting products like crude oil, natural gas, and petroleum products. The statistics for the United Kingdom and Australia are 20,000 and 32,000 km, respectively. More than half of the US oil and gas pipeline network is over 40 years old and corrosion has caused 23% and 39% of failures of oil and gas pipelines, respectively (Anon, 2002).

Twenty percent of Russia's oil and gas system is almost at the end of its design life and it is expected that in 15 years time, 50% of their pipelines will be at the end of their design life (Mahmoodian and Li, 2017).

In Canada, there are 34,000 km of oil pipelines and 26,000 km of gas pipelines where the prevention of corrosion-related failures at reasonable costs is also the main concern (Sinha and Pandey, 2002).

"The Water Infrastructure Network" reported the annual cost for maintenance and operation of the US national drinking water system at US$38.5 billion per year, which includes corrosion costs of US$19.25 billion (WIN, 2000). A study undertaken by The Water Services Association of Australia reveals that aggregated annual corrosion cost to the Australian urban water industry is approximately US$736 million (WSAA, 2009).

In the United Kingdom there are approximately 310,000 km of sewer pipes with an estimated total asset value of £110 billion (OFWAT, 2002). The investment for repair and maintenance of this infrastructure is approximately £40 billion for the period of 1990−2015 (The Urban Waste Water Treatment Directive, 91/271/EEC, 2012). It has been known that sewer collapses are predominantly caused by the deterioration of the pipes. For cementitious sewers, sulfide corrosion is the primary cause of these collapses (Pomeroy, 1976; ASCE (69), 2007).

In Los Angeles, USA, approximately 10% of the sewer pipes are subject to significant sulfide corrosion, and the costs for the rehabilitation of these pipelines are roughly estimated at £325 million (Zhang et al., 2008). As an example of a European country, in Belgium, the cost of sulfide corrosion of sewers is estimated at £4 million per year, representing about 10% of total cost for wastewater collection and treatment systems (Vincke, 2002).

These statistics indicate that pipeline networks are faced with high emergency repair and renewal costs, and frequent charges arising from increasing rates of deterioration worldwide. On the other hand, budget limitations are significantly restricting pipeline networks and reducing their capabilities in terms of addressing these needs. Large investments are required for building new pipelines networks. It is unlikely to be able to replace the existing pipe infrastructure completely over a short period of time. Therefore to eliminate the high costs associated with pipeline failures, pipe network managers need to generate proactive asset management strategies and prioritize inspection, repair, and renewal needs of pipelines by utilizing reliability analysis. The failure assessment and reliability analysis of pipelines can help asset managers to provide an improved level of service and publicity, gain approval and funding for capital improvement projects, and manage operations and maintenance practices more efficiently (Grigg, 2003; Salman and Salem, 2012).

It is also well known that the consequence of the failure of pipelines can be socially, economically, and environmentally devastating, causing, for example, enormous disruption of daily life, massive costs for repair, widespread flooding, and then pollution. This warrants a thorough assessment of the likelihood of pipeline failures and their remaining safe life, which is the topic of this book.

1.2 Scopes of Pipeline Reliability Analysis

Accurate prediction of the service life of pipelines is essential to optimize strategies for maintenance and rehabilitation in the management of pipeline assets. Service life (of building component or material) is the period of time after installation during which all the properties exceed the minimum acceptable values when routinely maintained (ASTM E632-82, 1996).

The basis for making quantitative predictions of the service life of structures is to understand the mechanisms and kinetics of many degradation processes of the material whether it is steel, concrete or other materials. Material corrosion in pipeline networks is a matter of concern for both strength and serviceability functions. Loss of wall thickness through general corrosion affects the strength of the pipeline. To that effect, incorporating the effect of corrosion into the structural analysis of a pipeline is of paramount importance. There are several parameters which may affect corrosion rate and hence the reliability of pipelines. To consider uncertainties and data scarcity associated with these parameters, various researches on probabilistic assessment of buried pipes have been undertaken (De Belie et al., 2004; Sadiq et al., 2004; Davis et al., 2005; Salman and Salem, 2012; Mahmoodian and Alani, 2014; Mahmoodian and Li, 2017).

Since the deterioration of buried pipelines is uncertain over time, it should ideally be represented as a stochastic process. A stochastic process can be defined as a random function of time in which for any given point in time the value of the stochastic process is a random variable depending on some basic random variables. Therefore a robust method for reliability analysis and service life prediction of corrosion affected pipes should be a time-dependent probabilistic (i.e., stochastic) method which considers randomness of variables to involve uncertainties in a period of time.

In most of the literature, failure and reliability assessment of pipelines has been carried out by considering one failure mode (Davis et al., 2005; De Silva et al., 2006; Moglia et al., 2008; Yamini and Lence, 2009; Zhou, 2011). However in reality, even in simple cases composed of just one element, various failure modes, such as flexural failure, shear failure, buckling, deflection, etc., may exist. To have a more accurate reliability analysis and failure assessment, multifailure mode of pipelines is also explained in Chapter 3 as system reliability analysis.

For a comprehensive reliability analysis, evaluation of the contributions of various uncertain parameters associated with pipeline reliability can be carried out by using sensitivity analysis techniques. Sensitivity analysis is conducted as a main part of reliability analysis from which the effect of different variables on service life of pipelines can be investigated. Sensitivity analysis is the study of how the variation in the output of a model (numerical or otherwise) can be apportioned, qualitatively or quantitatively, to different sources of variation (Saltelli et al., 2004; Mahmoodian and Alani, 2016). Among the reasons for using sensitivity analysis are:

- To identify the factors that have the most influence on reliability of the pipeline.
- To identify factors that may need more research to improve confidence in the analysis.
- To identify factors that are insignificant to the reliability analysis and can be eliminated from further analysis.
- To identify which, if any, factors or groups of factors interact with each other.

With a focus on "structural reliability", factors affecting both the structure and function of pipelines are analyzed and these are predicated on material type, design methods, and uses. As structural reliability is defined as a multifaceted term which considers factors that either have a direct or partial impact, deterioration mechanisms are also analyzed to illustrate the degree of impact on pipelines under various conditions.

Because corrosion is an ongoing issue, assessment of structural reliability of pipelines can also help to gain an insight into corrosion prevention mechanisms, by altering the unfavorable factors that contribute to this type of deterioration and consequently the service life of pipelines can be prolonged.

1.3 Types of Pipelines

Pipelines are tubular structures, typically underground and constructed from metals, plastics, and/or concrete for the purpose of carrying flow, including liquids and gasses.

The various types of pipelines are determined by the type of materials used to manufacture each pipeline based on its different applications (Table 1.1), the different functions it carries out, and the type of matter it is exposed to.

The engineering materials which can be used for pipe manufacturing are presented in Table 1.2. The commonly used pipe types are generally split into two categories: metals and nonmetals, with its subclassification: nonferrous/ferrous, plastics, ceramics, and composites.

Typically, metallic piping is derived from steel or iron, including unfinished black steel, carbon steel, stainless steel, galvanized steel, brass, and ductile iron.

TABLE 1.1 The Different Types of Pipe Material Based on Their Usage

	Usage		
	Wastewater Systems	**Drinking Water Supply Systems**	**Oil and Gas Supply Systems**
Types of pipe material	Cast iron (CI)	Galvanized steel (GS)	Steel
	Steel	Iron	Copper
	Galvanized iron (GI)	Copper	Yellow brass
	Copper	Polybutylene	Ductile iron
	Plastic	Unplasticized polyvinylchloride (PVC)	Aluminum
	Polyethylene (PE)	Chlorinated polyvinylchloride (CPVC)	Unplasticized polyvinylchloride (PVC)
	Unplasticized polyvinylchloride (PVC)	Polyethylene (PE)	Polyethylene (PE)
	Asbestos cement (AC)		
	Concrete		

Most of today's plumbing supply pipes are derived from steel, copper, and plastic, while pipes conveying wastewater are derived from steel, copper, plastic, and cast iron. Iron-based piping however, is subject to corrosion when used in highly oxygenated environments. Plastic pipes, including chlorinated polyvinyl chloride (CPVC) used in water supply systems is similar in polyvinyl chloride (CPV) pipes, which are used in waste lines, specifically lawn irrigation systems, except they differ in structural strength. Unlike CPV pipes, CPVC pipes do not soften when used in hot water streams. Furthermore CPVC is less expensive than copper by 15%−25%, making it ideal for use in water supplies.

The type of pipe used depends greatly on availability, its use, and the material it will carry. Most water supply pipes that deal with plumbing systems do not require complexity, such as steel pipes which are used inside buildings and are provided with zinc coatings. Other uses of steel pipes are water mains, sewerage systems, industrial water lines, plant piping, deep tube wells, casting pipes, and pipelines for natural gas.

Along with steel pipes, cement/concrete pipes are also widely used in sewerage systems. Although clay has been widely used in the past, it is not now used as frequently as steel and newer versions of concrete pipes. Clay pipes had an advantage over the other pipes in terms of strength, durability, and service life, as they were known to resist corrosion. However, they were prone to the attachment of tree roots on the surface of the pipes, causing cracking and ultimate failure.

TABLE 1.2 Engineering Materials in Which Pipelines Are Made From or Consist of/Include

Metals	Ferrous	Pure Ferrous	Wrought Iron
		Ferrous alloys	Iron alloys, cast Iron
		Steel	Nonalloy steels
			Alloy steels
	Nonferrous	Nonferrous alloys	Brass
			Bronze
			Solder
		Pure nonferrous	Aluminum
			Copper
			Zinc
Nonmetals	Ceramics	Clay	
		Glasses	
		Nanomaterials	
	Plastics	Elastomers	Rubbers
			Silicones
			Polyurethanes
		Thermoplastics	PVC
			Polyethylene
			Polypropylene
		Thermosets	Epoxies
			Polyimides
			Phenolics
	Composites	Concrete	
		Reinforced Plastics	
		Metal matrix	
		Ceramic matrix	
		Laminates	

New versions of concrete pipes, however, are also not resistant to root attachment. It has been reported that concrete sewer pipes might last in the long term, if they do not fail due to cracking or due to sulfur-based corrosion in sewer systems (Ridgers et al., 2006). Therefore, due to high strength and capacity of steel pipes, as compared to concrete and clay pipes, steel pipes are often used for pipe replacements when concrete and/or clay pipes fail.

The different types of pipes, along with their use, function, and viability status are reviewed in the following sections.

1.3.1 METALLIC PIPELINES

1.3.1.1 Nonferrous Pipes

Nonferrous metals, including aluminum, nickel, lead, tin, brass, silver, and zinc, are known for their tensile strength and present characteristics that hold an advantage over ferrous metals, mainly by their malleability, lighter weight, and corrosion resistivity. As iron content is absent in nonferrous metals, the probability of rust and corrosion occurring also is fairly low. Also these types of pipes can be used in pipe networks that are prone to highly corrosive environments, such as carrying corrosive chemicals and sewerage. Therefore, usage of nonferrous metals in water pipes and gutters are very common. However, the only difference lies in accessibility of these materials, recycling efficiency, and costs.

Recycling of nonferrous metals, including aluminum, copper, brass, and lead, is relatively common due to its scarcity. Therefore, there is a heavy dependence on recycling of scrap materials in order to create new nonferrous metals. In fact, aluminum is known to be the most recycled metal, ranking third in the world. It is also this reason that the cost of nonferrous metals is also more expensive than ferrous metals, with stainless steel often being used instead, due to its cheaper cost and versatility. Consequently, nonferrous metals are mainly used in fittings of pipes, and other small applications that require strength where needed.

Nonferrous materials, can be combined to create alloys, including brass, bronze, and solder, for increased strength and flexibility without resulting in the effects of heavier weight compared with ferrous metals. The high malleable state of nonferrous metals, makes them ideal for usage as they can be easily pressed into thin sheets without breaking.

1.3.1.2 Ferrous Pipes

Ferrous pipes are derived from ferrous metals consisting of iron. Different types of ferrous metals exist for different uses and are differentiated by the iron content percentage. Below is a list of the different types of ferrous metals:

- Mild steel: containing an iron content of 99.7%−99.9% and carbon content of 0.1%−0.3%.

- Stainless steel: Containing iron, nickel, and chromium content of various percentages, depending on the type of stainless steel. This type of metal is widely used as it is stain and corrosion resistant.
- Cast iron: containing 2%−6% carbon and 94%−98% iron content. Generally is used for manhole covers and sewer pipes, this metal is very strong but brittle.
- Wrought iron: composed of almost 100% iron

Ferrous metals are known for their tensile strength and durability and their characteristics recognized by their use in tall buildings and long bridges. They also have magnetic properties, and their element content includes carbon, nickel, and chromium. Ferrous metals are widely used in piping. Although high amounts of carbon causes most of these metals an increased proneness to rust and corrosion, metals containing wrought iron (lack of sulfur and carbon) and chromium content can resist oxidation and thereby provide protection against corrosion. It is an understatement however to suggest that wrought iron does not corrode. The lack of sulfur and carbon in wrought iron suggests that the iron is so pure that it resists oxidation, which is a necessary process for corrosion to occur, however, the wrought in iron only slows down corrosion and does not fully prevent it. In fact, it is known that most ferrous metals are associated with rust and corrosion.

Considering both the characteristics of ferrous and nonferrous metals and their uses in pipelines, it can be understood that the type of metals used in pipes depends on the function of the pipe itself. Ferrous metals would be used in pipes that are large and require sturdiness, due to their heavier weight, whereas nonferrous metals would be used in pipes that require flexibility, light weight, and are corrosion prone (Alton Materials, 2017).

Galvanized Iron Pipes

Galvanized iron pipes are wrought steel pipes which are often provided with a zinc coating. They are applied to plumbing systems inside buildings and are used in water supply and sewerage systems. These types of pipes are connected with jointing methods using T-connectors branching off the main line and Y-joints to split a single line into two and connectors to join the ends of pipes. The jointing methods include screwed joints, welded joints, grooved joints, and flanged joints. The fittings of these pipes are usually standardized, available in light, medium, and heavy grades depending on the thickness of the metal. The sizes range from DN 20 mm−DN 300 mm which are supplied in standard lengths of 5.8−6 meters.

The advantages of these pipes include, long life span, toughness, durability, availability in large diameters, and low installation and maintenance cost, which make them useful for large construction projects. They are also rust resistant, which makes these pipes very common in commercial buildings and households.

However, the handling of these pipes is particularly difficult as they are heavy as well as prone to blockages and difficult to repair when damaged. Although during the installation process, Teflon tape is used for fittings and the material is joined with copper to rule out potential leakage, damage is still not prevented. Furthermore to add extra support of horizontal hanging pipes, supports are placed every 6−8 ft, however certain factors including pressure changes, varying temperatures, and differing loads can impact these pipes, exceeding their strength and ultimately leading to failure.

Steel Pipes

Steel pipes are the most commonly used pipes in water supply systems. They are also used in pipelines for natural gas, and sewerage systems. Although comparatively expensive to other pipes, they hold the advantage of being able to withstand high pressures and are available in more convenient lengths, and can also be welded easily, thereby resulting in lower installation and transportation costs. These types of pipes are highly efficient and can be used in small diameters as needed and are 100% recyclable compared to other materials. The pipes can further be melted down and turned into other usable material in industry. Furthermore, the high strength of these pipes and resistance to damage caused by human errors, tree roots, and extreme weather conditions make these pipes the ideal choice for most water and sewerage supply systems.

The disadvantages of steel pipes include thermal conductivity, which is very poor as there is a difference in heat transfer. These types of pipes are usually bonded with aluminum or copper to increase thermal conductivity and improve heat transfer. Cost is another issue, as these pipes are expensive and this is guided by the misconception of being a one-time purchase. However, steel pipes are difficult to fabricate and lack the malleable qualities that other materials have, therefore repairs and replacements of steel pipes are extra difficult.

Seamless pipes

Seamless pipes are derived from solid steel that is in sheet or bar form and is formed into a solid round shape known as "billets" which are then heated and cast over a form such as a piercing rod to create a hollow tube or shell. These kinds of pipes are known for their ability to withstand pressure more efficiently in comparison to other methods of pipe manufacturing processes, as well as being fast and cost-effective. Seamless pipes are generally used in gas lines, as well as pipes that carry liquids (JSTEEL, 2017).

Because seamless pipes are able to withstand high pressures, they are also widely used in high-pressure applications including refineries, hydraulic cylinders, hydrocarbon industries, and in Oil and Gas infrastructure (Pearlite Steel, 2017).

In comparison to other types of piping, seamless pipes do not require any welding or joints and are simply formed by solid round billets, which adds on to its strength and other characteristics including corrosion resistance. According to ASME, these pipes are also more effective in withstanding mechanical stress and have a higher operating pressure than welded pipes which are known as nonseamless pipes (Ferrostaal, 2017).

In general, the application of seamless pipes depends on the thickness of the pipe wall. Higher temperatures are required to produce pipes with thicker walls which reduces deformation resistance resulting in a larger deflection (XINLIN, 2014).

1.3.2 NONMETALLIC PIPES

1.3.2.1 Plastic Pipes

These types of pipes are used in external and internal plumbing systems to carry cold water. Specifically, rigid PVC pipes are used to carry water within temperatures below 45°C as higher temperatures decrease the strength of these pipes. Sunlight and frequent changes in temperature also reduces the life of PVC pipes.

The three common types of plastic pipes are outlined below:

- Unplasticized PVC (UPVC) or rigid pipes, which are used in cold water systems.
- Plasticized PVC pipes, which include the addition of rubber, and have a lower strength and carry water of lower temperatures than UPVC pipes.
- Chlorinated PVC (CPVC) pipes, which are used in hot water systems, with temperatures of up to 120°C.

These types of pipes are joined using solvent cementing and heat fusion. The high durability as well as the capacity to function as insulation and resistance to corrosion make these pipes very usable in plumbing systems. The downfall of these pipes however are the temperature limitations, especially heat, as well as their weight and cost. These types of pipes are heavy to handle and are expensive to replace.

CPVC pipes are known to be widely used in hot and cold water systems. However, the function of CPVC pipes is compromised if contamination were to occur. A study conducted on the use of CPVC pipes in fire sprinkler systems observed that the potential causes of failure were more than one, with the most common cause of failure being contamination. The contamination was reported to occur inside the pipes and fittings when exposed to chemicals that were not compatible with the use of these particular type of pipe. Although it was not the exposure to harsh chemicals alone that caused failure, but rather the combined affect of stresses with incompatible chemicals. This action resulted in environmental stress cracking (ESC) and it is understood that due to the sprinkler system being static, the contaminants were sucked into the CPVC piping system due to pressurization and accumulated there.

Common issues that contribute to failure in these types of pipes are poor installation methods and manufacturing processes. Before installation, the pipe walls remain in a "frozen" state with a low level of stress, and during usage stresses within these pipes increase resulting in a requirement to release the stress. Exposure to heat, organic matter, and other chemicals including hydrocarbons allow the structural composition of the pipe to relax, by absorbing these materials, and causing softening of the pipe surface. This results in the overall softening of the pipe and can lead to longitudinal hairline cracking.

Causing additional stress on the pipe during installation can increase sensitivity towards ESC. Furthermore, temperature variances occurring inside these pipes can also contribute to ESC, by causing changes in pressure, therefore increasing the likelihood of more contaminants that can cause cracking and ultimately failure within CPVC pipes.

Thermoplastic pipes

Propylene random copolymer (PRP) pipes: The PRP pipe is one of the latest pipes and is used in cold and hot water systems. These pipes function both in sanitary and pure water pipelines as well as underground heating systems and hot water recycling systems. They are also useful in carrying compressed air, industrial water, and chemical materials. PRP pipes are considered to be the optimal pipe material for hot and cold water systems due to their high functionality, safety, cost-effectiveness, and high lifetime of up to 50 years. This type of pipe does not follow the aging process of most pipes and time-based leakage is almost nonexistent once installed and pressure tests have proven sufficient. The smooth inner wall and structural stability, as well as the capacity to function as heat insulation make this type of pipe reliable in supporting materials of various chemical compositions, pressures, and temperatures.

The disadvantages of these particular pipes, include the increased need for higher technical requirements, including the usage of special tools and professionals in the installation process to ensure safety of the systems involved. Another factor is that companies involved in construction do not fully rely on these pipes, therefore they are not widely used as they are not as popular as other materials used in pipeline industry.

High Density Polyethylene (HDPE) pipes: HDPE pipes are made from polyethylene thermoplastics derived from petroleum through a heating process, where petroleum is exposed to high temperatures to form ethylene gas. These gas molecules then link together, forming a polymer chain, referred to as polyethylene (Kumar et al., 2011).

These types of pipes function as underground pipelines, carrying fresh drinking water, gases, corrosive substances, and sewerage. The easy installation process of these pipes, ensures that they do not require connectors to form a pipeline network compared to metal piping installation methods. The pipes are simply fused together, creating a monolithic system, resulting in an antileak HDPE pipeline.

Due to their strength, durability, flexibility, and light weight characteristics, HDPE pipes are ideally used as water and gas pipes. Their ability to resist high stresses, pressure, leakage, tearing, and corrosion makes HDPE pipes very useful both as buried and on-ground pipes. These pipes are also environmentally sustainable, as they are nontoxic, and chemical-resistant, therefore they are ideal for water application, energy, and industrial usage.

Because the surfaces of HDPE pipes are smoother than ferrous and concrete pipes, it ensures there is less drag in terms of flow, therefore these pipes hold great hydraulic characteristics, ensuring good volumetric flow rate, and less turbulence with higher flows. The chemical resistance of these pipes also eliminates any sticking of chemicals, bacteria, fungi, and other organic substances to the pipe surfaces thereby ensuring prolonged service life.

In terms of weather resistance and long service life, HDPE pipes have the capacity to withstand freezing water more efficiently than metal pipes and minimal heat transfer, making them excellent insulators. Even at temperatures exceeding 35°C, HDPE pipes do not become brittle and show lower friction loss rates than metal pipes. Furthermore, these pipes can also resist UV, if carbon black is added to the HDPE.

1.3.2.2 Composite Pipes

Concrete Pipes: Concrete pipes are used for drainage purposes, including storm water, road culverts, and sewers. Manufactured but not limited to 2.44 m lengths as well as being custom made, with specific pipe fittings, these types of pipes are usually buried under roads of urban cities, therefore it is important to ensure the structural safety and durability of these types of pipes. As concrete pipes are usually difficult to repair and replace, it is important to ensure that certain measures are taken to preserve their life span. These particular types of pipes are susceptible to sulfide-based corrosion, and their design must ensure adequate structural safety.

When considering storm water pipes and sewerage pipes, providing a high axial load transfer capacity and a flexible high watertight joint is critical to ensure safe installation and proper function.

The expected service life of drainage pipes is usually between 80 and 100 years, however factors including design methods, loads, installation errors, climate, and corrosion all affect the service life of these pipes. Installation errors such as those related to depth and width of burial are usually hard to be prevented. Also, factors such as climate change, which are going to continue having direct effect on the corrosion process of concrete pipes, can not be prevented easily (Tran, 2014)

Glass Reinforced Epoxy (GRE) pipes: GRE is essentially a glass fiber that forms into a polymer. This combination is known as a glass fiber-composite. This composite is then combined with epoxy resin to make a glass fiber-reinforced polymer matrix composite, with epoxy resin as a matrix. This combination is classified as glass reinforced epoxy. The epoxy resin is effective particularly for its high wetting power and easy adherence to the glass fiber as well as adequate cohesion strength (Asi, 2009).

Furthermore, epoxy resin does not shrink considerably when setting and also holds thermal properties. Therefore, using epoxy resin as a matrix in glass fiber composites increases the likelihood of producing a high strength and weather-resistant material.

Due to the extensive use of GRE in automobiles, marine, aerospace, and defense industries, it is evident that GRE is a reliable material, showing characteristics that are ideal for usage in piping. These characteristics include, corrosion and chemical resistance, light weight, and extensive service life.

GRE piping also has an efficient strength/stiffness ratio, suggesting that it provides a good balance if one was to outweigh the other due to external factors such as loading and weathering and internal factors including flow of materials and pressure changes. The high temperature resistance (above 125°C) and high electrical insulation of these pipes also allows use in hot water and industrial systems.

The ability of GRE pipes to resist thermal expansion also ensures high fatigue performance in the sense that the structural expansion of the pipe is minimal, therefore it will have a small effect on pressure changes, flow disruptions, and turbulences within the pipe which may exacerbate aging of the pipe.

1.3.2.3 Ceramic Pipes

Clay pipes: Clay pipes are generally under the term of "Vitrified" clay pipes, which is essentially the properties of clay being transformed into a glass-like substance via application of heat. This transformation makes vitrified clay pipes harder than steel in terms of structure, however they do not exceed the tensile strength of steel.

Clay pipes are known for their strength, longevity, and performance in terms of environmental sustainability. Their strength minimizes the probability of accidental damage during installation, and disturbance, including unexpected imposed loads, subjected to the pipes during traffic loads. This is supported by the fact that clay pipes are rigid, and are not particularly prone to flattening and deflection under loads.

One of the characteristics of clay pipes is the ability to maintain consistency in terms of structure and function due to its physical properties. This is further highlighted by the impermeability of clay pipes to the surrounding environment, including a reduction in the risk of the leakage of pipe contents into the surrounding soil. Therefore it is safe to suggest that vitrified clay can be resistant to aggressive ground conditions including those that contain chemicals and aggressive soils.

Furthermore, vitrified clay pipe was proven to be the only type of pipe that is resistant (over centuries of usage) to sulfide-based corrosion and corrosion triggered by aggressive ground conditions. It is safe to suggest that the vitrification process of a clay pipe has been proven chemically to act as a built-in protection mechanism against any deterioration and strength loss. This concludes that vitrified clay pipes do not need a protective coating for protection against corrosion, as these pipes have a long service life (Gladding McBean, 2017).

1.4 Design of Buried Pipelines

The design of buried pipes constitutes a wide ranging and complex field of engineering, which has been the subject of extensive study and research in the world for many years. There are two main stages for designing of pipes: (1) hydraulic design, and (2) structural design. In the hydraulic design stage, the focus is on determination of the demand of the system for collecting and/or conveying the flow. Based on this, the diameter of the pipe is estimated. In the second stage, focus is on determination of structural capacity or strength, including details like wall thickness and/or reinforcement. This section discusses the structural design of buried pipes. It introduces and compares different existing design methods. The structural properties of the pipe are analyzed to ensure the pipe can safely sustain external and internal loads during its service lifetime, without loss of its function and without detriment to the environment.

A set of performance criteria must be met when the pipe is subjected to loads. As for other structures, there are two categories of performance criteria for underground pipes: ultimate limit state, and serviceability limit state.

The ultimate limit state is represented by the strength of the pipe and is reached when the pipe collapses or fails in general. Flexural and shear failures are the two main ultimate limit states that are considered in design and assessment of pipelines (ASCE 15-98, 2000). Serviceability limit states may be measured by cracking or other functional requirements (e.g., leakage, deformation beyond allowable limits (for flexible pipes), and excessive movement at the joints).

The principle for the design of a pipe is to ensure that both serviceability and ultimate limit states are not reached. This includes consideration of one or more of the following conditions: strain, stress, bending moment, and normal force or load-bearing capacity, in the ring or longitudinal direction as appropriate; and water tightness.

The design of a buried pipe involves the selection of appropriate pipe strength and a bedding combination which is able to sustain the most adverse permanent and transient loads to which the pipeline will be subjected over its design life.

The design of buried pipelines depends on factors including the structural properties of particular pipes, the internal and external pressures, loads, and the condition of the surrounding soil. The general steps involved in the design of buried pipelines are outlined below:

Step 1: Determining the wall thickness of the pipe due to internal pressure. The wall thickness depends on the diameter of the pipe, flow pressure, and routing of the material inside the pipe.
Step 2: Checking minimum wall thickness of the pipe for handling.
Step 3: The external loading during construction is analyzed in the pipe–soil embedment system, which includes earth loads, live loads, the water table conditions, and the pressure in the pipe.

One of the most crucial factors to consider in pipe design is ring theory, which stems from external loadings. External loadings will be analyzed to highlight the stability of the pipe ring and it is dependent on the properties of the surrounding soil, as soil–pipe interaction is known to have an effect on the structural behavior of the pipe. However, before understanding ring theory it is important to gain an understanding of the stresses in the supporting soil surrounding the pipe. Furthermore, the stiffness factor of the pipe is also a factor to consider, especially its sensitivity based on the time of loading and how it can be altered by temperature changes.

To illustrate pipe design methods of buried pipelines case scenarios are used where pipes are subjected to different conditions. Steel pipes are considered as examples to explain what factors need to be considered for pipe design.

Case 1: Determining wall thickness for internal pressure and handling
Thickness of steel pipe is determined by ensuring the maximum value for the internal pressure that is being analyzed does not exceed the limiting hoop tensile stress. Hoop tensile strength is determined using σ_h.

It can be seen in Fig. 1.1 that the illustrated cross-section of a steel pipe is in static equilibrium, with forces equating to:

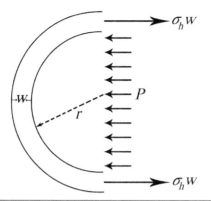

Figure 1.1 Cross-section of a pipe in static equilibrium.

$$2 \times \sigma_h w = P \times 2r \tag{1.1}$$

The above equation representing hoop tensile stress, ultimately results to:

$$\sigma_h = \frac{Pr}{w} \tag{1.2}$$

In order to determine pipe wall thickness, the formula is rewritten as:

$$w = \frac{PD_0}{2\sigma_P} \tag{1.3}$$

where

σ_P = stress from internal pressure (psi)
P = pressure (psi)
D_0 = steel cylinder outside diameter (inches)
w = thickness of steel pipe

The most common internal pressure that is analyzed for the design of pipelines is operating pressure, known as the working pressure (P_w).

The allowable hoop tensile stress of the steel pipe is typically limited to a value equal to 50% of the minimum yield strength of the pipe and is higher (75%) with transient pressures (Whidden, 2009). As this pressure is usually referred to as temporary due to its momentary spike, a greater hoop tensile stress is allowed. Following this design procedure, the pipes are hydrostatically tested to ensure factors such as tensile stresses do not exceed yield strengths and capacity of the steel.

The minimum thickness for steel pipes often depends on the safety of handling and installation. Typically, D/w ratios of up to 240 are used, however for pipes constructed with special design specifications, D/w ratios of up to 288 and higher can be used, which are generally seen in pipelines that are used in irrigation and hydroelectric systems.

Case 2: Ring stability—steel pipe exposed to external pressure, water table, and vacuum pressure

Ring stability of a pipe is determined by analyzing the amount of deflection required to cause soil slip with or without added vacuum pressure and water table. In most cases, external pressure is present and is always considered along with other given factors. In this case, the steel pipe is also exposed to vacuum pressure, therefore it is important to analyze the amount of vacuum pressure required to cause soil slip to prevent pipe failure.

However, ring stability of a pipe is not only limited to the above factors. Pipe stiffness also plays a significant role in the ring stability of pipes. Factors that can affect the ring stability of pipes are ring stresses, which, if exceeding pipe capacity, ultimately lead to pipe failure by causing pipes to collapse.

Ring stiffness of a pipe is defined as the ability to resist deflection. The stiffness of a pipe is measured via a ratio consisting of the applied load(s), F, over the measured deflection, D, as follows:

$$\text{Pipeline stiffness} = \frac{F}{D} \tag{1.4}$$

Ring stiffness however is determined via the equation below and in this case per unit length of pipe:

$$\text{Ring stiffness} = \frac{EI}{r^3} \tag{1.5}$$

As $I = \dfrac{w^3}{12}$ and $r = \dfrac{D}{2}$ then:

$$\text{Ring stiffness} = \frac{2}{3}E\left(\frac{w}{D}\right)^3 \tag{1.6}$$

Following installation of the pipe, it is important to also pressurize the pipe. This pressurization leads to the internal pressure inside the pipe decreasing deformation of the ring caused by deflection.

However, deflection of pipe does not exceed beyond the vertical compression of the surrounding soil. The vertical deflection, at which soil slip occurs, is referred to as the critical deflection of soil. When this soil slip occurs, it causes the pipe to collapse (if flexible).

To minimize this effect, cement mortar coatings and linings are added to pipes, although deflection limits can be better controlled by ensuring proper selection and management of embedment material, and installation techniques, instead of controlling pipe stiffness.

In order to control pipe stiffness, not only an increased number of factors need to be considered, which could prove to be complicated, but in the case of pipe failure it will be more difficult to pinpoint the exact cause of the failure. Although the addition of pipe coatings and linings can slow down pipe deformations, small cracks still tend to appear in these coatings and linings. However, these cracks are not critical, providing they are small, as moist environments close the cracks by the formation of calcium carbonate ($CaCO_3$) in the cement. The size of the cracks is determined by a comparison to the thickness of a dime, with cracks being no bigger than a dime considered small.

The mechanism, in which the cracks are closed, is due to the pressure in the pipe, which causes the pipe to deflect, forming a more rounded shape and ultimately "healing" the cracks.

However, low pressures and gravity flows inside pipes can result in permanent ring deformation of the pipe causing the cracks to expand in size. This can cause water exposure to the steel, thereby resulting in corrosion of the pipe.

Cracks wider in size are known to occur in the tensile zones of pipes, however in the case of coatings, the cracks typically occur at the spring line, as shown in Fig. 1.2. Cracks in the linings of pipes are known to occur at the crown and invert.

Cracks in pipes are determined by taking the width (δ) of the widest single crack and a calculation based on the location of the crack via the equations below (Whidden, 2009):

$$\frac{\delta}{2w_c} = \frac{1}{r_{min}} - \frac{1}{r} \text{ coating} \tag{1.7}$$

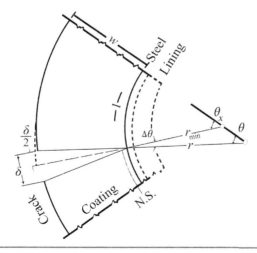

Figure 1.2　Crack width, δ, at the springline in a cross-section of a pipe.

$$\frac{\delta}{2w_l} = \frac{1}{r} - \frac{1}{r_{max}} \text{ lining} \tag{1.8}$$

where:

δ = width of the crack

w_c = thickness of mortar coating

w_l = thickness of mortar lining

r = circular radius of pipe

r_{max} = max radius

r_{min} = min radius

Cracks are usually narrow and their width does not exceed the thickness of a dime if the deflection of the pipe ring is less than 5%. As the pressure in the pipe results in a rerounding of the pipe, and causes the cracks to heal, those remaining are only hairline cracks which are distributed evenly throughout the pipe and heal in moist environments.

Despite this measure, the best method to determine crack width is not observing pipe ring deflection but rather focusing on permanent deviations in radii of curvature, which is a more reliable method.

A theory that involves a method on the basis of radii is longitudinal strain, which supports the formula below:

$$\frac{\sigma_z}{E} = \frac{r}{R} \tag{1.9}$$

where R is radius of longitudinal bend in pipe, r is radius of the pipe, and E is modulus of elasticity.

Fig. 1.3 shows the compound yield of a pipe, it is devised that the yield stress is slightly greater than the tensile strength, σ_y, if both the longitudinal and ring stresses hold the same sign (either tension or compression); conversely, the yield stress is less than tensile strength, σ_y, if the longitudinal and ring stresses hold opposite signs.

Although yield stress is known to be a measure of performance limit in the design of pipes, it is not always regarded as a failure condition, however in the case of steel pipes, yield stress is biaxial in the worst cases. Yield stress is measured via the application of standard unaxial tension tests. An example of this is shown in Fig. 1.3 which shows that steel pipes containing internal pressure, the hoop tension stress is σ_x, while longitudinal stress, σ_z, causes biaxial stress. Longitudinal stresses are associated with longitudinal bends in pipes, with tension and compression acting on the inside and outside of the bend, respectively. Therefore, it is safe to conclude that both hoop tension and longitudinal

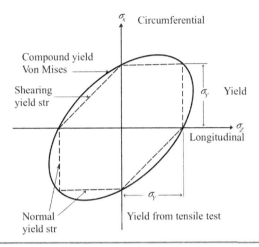

Figure 1.3   A compound yield of a pipe.

stresses are compound stresses that act on the pipe and alter the yield stress (Whidden, 2009).

The compression component in longitudinal stresses is related to external pressure being applied to the pipe, and the stress found in the pipe wall is defined as ring compression stress, measured by the formula below (Whidden, 2009):

$$S = \frac{P(D_O)}{2w} \tag{1.10}$$

External pressures can be caused by soil embedment (including water table height above ground surface) with additional pressures being taken into consideration as these pressures need to be supported by the embedment. In the case of external pressures including water table heights above ground surface, it is essential that the water table height is added to the internal vacuum pressure as the internal vacuum pressure is measured at the height of the water table. It is in this instance where soil slip occurs and it is important to ensure all external pressures are taken into account when designing the pipe ring.

Although the pipe does not exactly fail at yield stress, this measure is set as a performance limit where the performance limit of pipes is generally wall buckling or crushing of the pipe wall at yield stress, σ_y.

The ring compression design for pipes is determined via the formula below with the safety factor, sf (Whidden, 2009):

$$\frac{\sigma_y}{sf} = \frac{P(D_O)}{2w} \tag{1.11}$$

The internal pressure decreases ring formation within the pipe after installation and pressurization. By this, the deflection of the pipe is not greater than the vertical compression of the surrounding soil. However, once the critical deflection, which is the vertical deflection at which soil slip occurs, exceeds this vertical compression, the pipe will collapse (if flexible pipe).

The deflection of the pipe however, may be more limited than that for critical soil compression as coatings and linings in pipes may play a role in the process. Deflection can be controlled more by controlling the material in the embedment and the installation processes, than the properties of the pipe itself, such as pipe stiffness.

steel is ductile and is known to perform in a ductile range, the above factors should be considered carefully in pipe design, as steel pipes are subject to factors including bending, buckling, cracking, and collapsing that significantly affect their structural integrity. The next section outlines how these factors contribute and result in damage and failure mechanisms.

1.5 Loads and Stresses on Pipelines

1.5.1 LOADS ON PIPELINES

All pipes shall be designed to withstand the various external and internal loadings to which they are expected to be subjected, during construction and operation. The external loadings include loads due to the backfill, most severe surface surcharge or traffic loading (live load) likely to occur, and self-weight of the pipe and water weight. The internal pressure in the pipeline, if different from atmospheric, shall also be treated as a loading.

1.5.1.1 Earth Load

Beginning in 1910, Marston and Anderson developed a method for calculating earth loads above a buried pipe based on the understanding of soil mechanics at that time. Marston's formula is considered for calculation of earth load on buried pipes in codes of practice and manuals (such as BS EN 1295-1 and ASCE No.60). The general form of Marston's equation is:

$$W = C\gamma B^2 \tag{1.12}$$

where W is the vertical load per unit length acting on the pipe because of gravity soil loads, γ is the unit weight of soil, B is the trench width or pipe width, depending on installation condition, and C is a dimensionless load coefficient depending on soil and installation type (available in design manuals).

The pressure distribution around the pipe from the applied loads (W) and bedding reaction shall be determined from a soil-structure analysis or a rational approximation. Acceptable pressure distribution diagrams from soil-structure analysis are the Heger Pressure Distribution (Fig. 1.4A) for use with the Standard Installations, the Olander/Modified Olander Radial Pressure Distribution (Fig. 1.4B), or the Paris/Manual Uniform Pressure Distribution (Fig. 1.4C).

1.5.1.2 Pipe and Flow Dead Loads

The dead load of the pipe weight shall be considered in the design based on material density. The dead load of fluid in the pipe also shall be based on the unit weight of the stream (ASCE 15-98, 2000).

1.5.1.3 Live Load

In designing buried pipes, it is necessary to consider the impact of live loads (surcharge) as well as the dead loads. Live loads become a greater consideration when a pipe is installed with shallow cover under an unsurfaced road way, railroads, and/or airport runways and taxiways. Surcharge loads are calculated using Boussinesq's theory (Moser and Folkman, 2008), for various vehicle wheel loading patterns, representing the most severe loadings which might apply in various locations.

Both concentrated and distributed superimposed live loads should be considered in the structural design of sewers. The following equation for determining loads due to superimposed concentrated load, such as a truck wheel load has been presented by ASCE No.60 (2007):

$$W_{sc} = \frac{C_s P F}{L} \tag{1.13}$$

where

W_{sc} = the live load on the sewer in kg/m of length
P = the concentrated load (kg)
F = the impact factor
C_s = the load coefficient, a function of $\frac{B_c}{2H}$ and $\frac{L}{2H}$ where H is the height of fill from the top of pipe to ground surface in (m) and B_c is the width of the sewer in (m)
L = the effective length of sewer in (m)

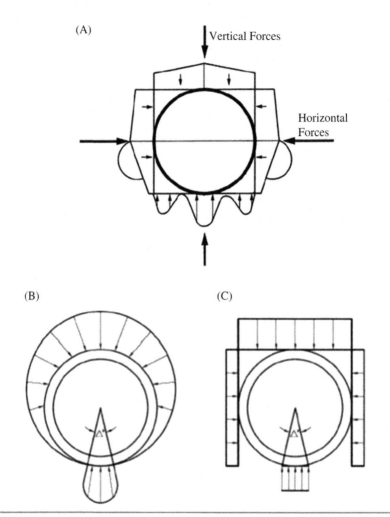

Figure 1.4 (A) Heger earth pressure distribution; (B) Olander/Modified Olander radian pressure distribution; (C) Paris/Manual uniform pressure distribution (ACPA, 1993).

For the case of superimposed load distributed over an area of considerable extent, the formula for load on pipe is (ASCE No.60, 2007):

$$W_{sd} = C_s p F B_c \qquad (1.14)$$

where

W_{sd} = the load on the pipe (N/m)
p = the intensity of distributed load (N/m^2)

F = impact factor

B_c = the width of the sewer pipe (m)

C_s = the load coefficient, which is a function of $\frac{D}{2H}$ and $\frac{M}{2H}$, and D and M are width and length, respectively, of the area over which the distributed load acts

H = height from the top of the sewer to ground surface (m)

1.5.2 STRESS IN BURIED PIPELINES

Rajani et al. (2000) developed a formulation for total external stresses including all circumferential and axial stresses in buried pipelines. σ_θ is hoop or circumferential stress:

$$\sigma_\theta = \sigma_F + \sigma_S + \sigma_L + \sigma_V \tag{1.15}$$

where σ_F is hoop stress due to internal fluid pressure, σ_S is soil pressure, σ_L is frost pressure, and σ_V is traffic stress.

Similarly axial stress, σ_x, would be:

$$\sigma_x = \sigma_{Te} + \sigma_{\dot{F}} + (\sigma_S + \sigma_L + \sigma_V)\nu_p \tag{1.16}$$

where σ_{T_e} is stress related to temperature difference, $\sigma_{\dot{F}}$ is axial stress due to internal fluid pressure, ν_p is pipe material Poisson's ratio, and other parameters have already mentioned. Equations and references used for the abovementioned stresses have been presented in Table 1.3.

The factors that impact the structural integrity of pipes, mentioned above, are also in relation to performance limits. Once these factors exceed the structural

TABLE 1.3 Stresses on Buried Pipes

Stress Type	Model[a]	References
σ_F, hoop stress due to internal fluid pressure	$\dfrac{pD}{2d}$	Rajani et al. (2000)
σ_S, soil pressure	$\dfrac{3K_m\gamma B_d^2 C_d E_p dD}{E_p d^3 + 3K_d p D^3}$	Ahammed and Melchers (1994)
σ_L, frost pressure	$f_{frost}.\sigma_S$	Rajani et al. (2000)
σ_V, traffic stress	$\dfrac{3K_m I_c C_t F E_p dD}{A\left(E_p d^3 + 3K_d p D^3\right)}$	Ahammed and Melchers (1994)
σ_{T_e}, thermal stress	$-E_p \alpha_p \Delta T_e$	Rajani et al. (2000)
σ_P, axial stress due to internal fluid pressure	$\dfrac{p}{2}\left(\dfrac{D}{d} - 1\right)\nu_p$	Rajani et al. (2000)

[a]Notations introduced in Table 5.6, Chapter 5.

capacity of pipes, it can lead to damage and failure, presented in the form of "performance limits" which include leaks, ruptures, erosion, stress corrosion cracking issues (SCC), and fatigue.

Before understanding these performance limits it is important to recognize the difference between what factors simply contribute to performance limits and what factors cause and present an end result of performance limits. An example of this would be the loads and stresses on pipes which define the conditions for performance limits which will contribute to ring or longitudinal deformation of pipes. As a result, these deformations will eventually lead to cracks, leaks, ruptures, and collapsing of pipes, which are defined as performance limits.

However before gaining an understanding of this concept, it is useful to identify that this logical idea actually lies within the pipe–soil interaction mechanism. The pipe–soil interaction mechanism is essential as soil is not only considered as a load on pipes, but it is a component of the pipe structural system. Therefore, this mechanism is what defines the factors that contribute to performance limits and ultimately what causes them.

The structural performance depends on the pipe–soil interaction. In this mechanism, the properties of soil play a significant role in keeping the pipe in shape and in place. Both the stiffness of the soil and the pipe ring are defined as resistance to deflection with soil stiffness defining most of the resistance to deflection of flexible steel pipes and ring stiffness ensuring the pipe maintains its shape during the implementation process of the embankment and compaction. However, ring stability of pipes can undergo spontaneous deformation causing instability, which becomes a performance limit. This ring instability can then cause pipes to invert if the deflection of the pipe ring and soil slippage occurs simultaneously, causing a change in structure and function of the pipe. Therefore, the mechanism by which ring stability is maintained is through the pipe–soil interaction.

As pipes are subject to external loads, surface loads, construction, and hydrostatic and soil pressures, having a required minimum soil cover is critical. The minimum soil cover is important for preventing damage from the external loads as well as damage from weather conditions. To further explore this issue, the stress and strain relationship in the design of buried flexible pipes is analyzed. It is understood that soil compression is defined as soil strain and in excess, causes the shifting and deflection of pipes. To prevent this action, it is essential that the embedment does not exceed the compression limit as it results in nonuniform bedding. This causes adjoining pipes to deflect at a joint, resulting in leakage and possibly rupture. The excessive compression also causes a flattening at the bottom of the pipe, which alters the ring deflection ratio (Δ/D) of the flexible steel pipe. This ratio states that the ring deflection of the pipe is almost equal to the vertical strain of the side-fill embedment. Given that strain is a function of stress and its relationship is nonelastic, it is observed that in pipes subject to any stress, compaction of the embedment is also a component that affects vertical strain, and thereby plays a role in the ring deflection ratio. It is observed that vertical strain decreases as compaction of the embedment increases. However, although it seems logical to implement a highly compacted embedment to maintain a stable ratio, it is also important to consider that improper embedments such as those with high strengths can lead to damage and failure of pipes. In the case of soil movement, high strength embedments are prone to cracks, leaving pipes to be the concentrated target of high stresses, and with higher stresses leading to a larger deflection, exceedance of allowable limits, and finally failure. Ultimately it is a cycle that continues, with one factor affecting the other.

The usage of proper embedment is an important concept that affects the pipe—soil interaction by ensuring the surrounding environment of the pipe will not cause any additional stress and strain that can impair the structure of the pipe. One of the issues that causes pipes to buckle and collapse is the misalignment of pipes in soils. The embedment ensures that the pipe is kept aligned in place on the bedding and protected from external loads. However, issues arise with soils subjected to improper fill materials in the embedment, high loads that exceed allowable limits, compaction issues, soil permeability, high water tables, seepage, creep, and temperature fluctuations.

Another major issue caused by loose soils, high water tables, saturation of soils under the water table, and external loads, particularly vibrations from traffic, is the liquefaction of soils. The "liquefied" state of soils can cause the pipe to float out of alignment and cause the pipe to collapse. This becomes very easy as liquefied soil unit weight is twice the unit weight of water.

Water seepage into soils can also cause soils to liquefy and resemble quicksand which causes flotation of pipes and the liquefaction of the soil embedment. Furthermore, the soil under the pipe can be washed out through groundwater flow channels caused by the seepage, thereby resulting in a significant change within the overall structure of the affected pipe. This is especially common in areas where temperature fluctuations and water content constantly changes, which present a "freeze—thaw" action, causing a "creep" in the surface layers of the soil. These conditions allow ground to landslide and greatly affect the pipelines underneath.

Due to the inevitable factors such as those mentioned above, liquefaction of soils may not be entirely preventable, although the risk of soil embedment becoming liquefied can be minimized in order to protect pipelines.

Ensuring that the correct material is used in the embedment can minimize the risk of factors that may cause the pipes to leak or collapse. It is important to ensure that the fill in the embedment is "flowable" and that the granular material which is used for fill purposes is well-graded. Large rock particles are generally avoided to prevent slippage, and a nonuniform surface, while fine rock particles are used, generally being less than 12% of total embedment material, to allow flowability as well as to ensure the fill is not too liquidated. For a flowable fill, Portland cement or fly ash is typically used and formed into a slurry which flows under the pipe, allowing a more uniform bedding and embedment.

It is essential that fill material is compacted with specific moisture content based on the depth of the pipe burial. Efficient compaction ensures the usage of well graded soils and can prevent piping, which is defined as the washing out of soil by groundwater flow channels from underneath pipes. Well graded soils also serve as a filter, preventing fine particles in the embedment from leaking out into the trench walls and vice versa which can lead to instability of soils.

It is expected that the embedment should be able to support live loads above pipes as well as the load of the backfill. To enable this, it is crucial that the pipe is protected by the embedment, by forming an arch over the pipe. To maintain this protective arch, the vertical compression of the soil needs to be controlled by keeping the compressive strength to a low minimum of 40 psi and ensuring it does not exceed the internal pressure of the pipe (Whidden, 2009).

As mentioned earlier, the deflection of the pipe ring is almost equal to the vertical compression of the side soil. Therefore, it is important that the vertical compression and shrinkage of the flowable fill which is placed as a side fill, does not exceed the allowable limits of pipe deflection. Excessive deflection can occur if the flowable fill fails to maintain the shape of the pipe and fails to support backfill. Therefore flowable fill is also required to hold an adequate bearing capacity to ensure loads are supported and the risk for stress—strain induced pipe deformation is significantly reduced.

To summarize in the context of the pipe—soil interaction mechanism, the performance limits are excessive compression and slippage of soils caused by the liquefaction of soils, mixing of soil particles, caused by groundwater seepage which contributes to soil instability, and the permeability of soils which affect water seepage by further accelerating the problem.

Soils that are liquefied and are subject to movement, can cause an increased susceptibility to larger strains that can exceed performance limits and cause additional stress which feeds into the failure mechanism of pipelines.

However, the liquefaction of soils caused by water seepage is not only limited to causing pipe failure directly via misalignment. The exposure of flexible metal pipes to water seepage can also contribute to corrosion and cause corrosion-induced failure in pipes.

1.6 Deterioration of Pipes

It is understood that different types of pipes undergo different types of deterioration, due to their varying structural and chemical composition as well as various environmental exposure. In this section different types of pipes and their predominant deterioration mechanisms are explained.

1.6.1 DETERIORATION OF CONCRETE PIPES

Structural deterioration of concrete pipelines is due to corrosion affecting the concrete itself by altering the structure of the pipe, including the formation of cracks as well as affecting the reinforcement inside. The corrosion of concrete is caused by the action of sulfuric acid, which reacts with the exposed concrete to form gypsum ($CaSO_4 \cdot 2H_2O$). The soft and soluble characteristics of gypsum allows it to penetrate the concrete matrix and work to gradually deplete the thickness of the concrete (Khalifeh et al., 2017).

This mechanism also works in the same manner in microbial-induced corrosion (MIC), which is also common in concrete sewer pipes. The bacterial action generates sulfate ions and is typically known to occur in sewer systems as these systems rely heavily on concrete pipelines. Concrete piping around aggressive soils, which can contain sulfate ions, is particularly prone to corrosion as the ions generated are the components of sulfur-based corrosion, and can act to further accelerate the corrosion process. The presence of chlorides and other caustic agents contributing to crack formation in concrete pipes can also further deteriorate the concrete and reach the reinforcing steel inside. Corrosion of the reinforcement results in complete dysfunction of concrete pipes as crack formation is so advanced by the time it reaches the reinforcement, and already shows signs of leakage, and localized failure.

Furthermore, pressure formation inside the reinforcement caused by the growth of cracks results in expansion of the reinforcement inside concrete pipes. This causes failure by the gradual cracking of the surrounding concrete.

Crack formation in concrete pipes generally occurs in cases where tensile stress exceeds tensile strength. As a result, a decrease in strength and capacity is observed in concrete pipes which show signs of sulfide-based corrosion. In the design of concrete pipes, it is important that load-bearing capacity is greater than the applied total static and dynamic loads. Soil stress around pipes and factors affecting the performance of the pipe should also be considered, including those relating to bending failure, shear failure, loss of concrete cover and excessive crack width of the concrete (Tran, 2014).

Below is a detailed explanation of the two deterioration mechanisms of concrete pipes, including reinforcement corrosion and microbial-induced corrosion.

1.6.1.1 Reinforcement Corrosion

The most causes of reinforcement corrosion are chloride intrusion, exposure to water and oxygen, and reduction in pH, which are the result of concrete pipes being subjected to corrosive chemical environments, such as petrochemical plants and sewers. Improper placement of reinforcement and concrete, including insufficient concrete coverage of steel reinforcement can heighten this effect and create a perfect ground for deterioration processes to take place. This can result in corrosion and expansion of the reinforcing steel, which can lead to cracks, and spalling in the concrete, ultimately causing a reduction in structural capacity.

To increase the tensile strength of concrete, steel reinforcements are used, however when tensile strength is exceeded by tensile forces acting on the concrete, it can result in cracking. The reinforcements also function to control the width of the crack, and prevent failure, however for this to occur, design capacity must be met.

Mechanism of Reinforcement Corrosion

The process of corrosion in reinforcing steel is set by two distinct reactions, the first being oxidation and the second being reduction. The product formed, as shown in the corrosion mechanism below, further reacts and produces rust on the surface of the steel bar.

The corrosion mechanism of steel reinforcement is illustrated in following reactions:

$Fe \rightarrow Fe^{2+} + 2e^{-}$ (anodic)
$2 H_2O + O_2 + 4 e^{-} \rightarrow 4 OH^{-}$ (Cathodic-reduction in oxygen)
$2 Fe^{2+} + 4 OH^{-} \rightarrow 2 Fe(OH)_2$ (chemical)(essentially a by-product for further reaction to form rust)
$$\downarrow$$
$4 Fe(OH_2) + O_2 \rightarrow 2H_2O + 2Fe_2O_3.H_2O$ (formation of rust)

The corrosion mechanism illustrated above, applies to the different types of corrosion that occur in the reinforcing steels of concrete pipes, including uniform, crevice, and galvanic corrosions.

Carbonation

This process is driven by diffusion, where carbon dioxide from the atmosphere diffuses through the porous concrete, reacting with water and calcium carbonate found in Portlandite, which is an oxide mineral occurring in the form calcium hydroxide and used in the making of cement. This process, illustrated by the reaction below, acts to reduce the alkanity of the concrete to a pH of $8-9$, and thereby alters the stability of the oxide film.

$$Ca(OH)_2 + CO_2 CaCO_3 + H_2O$$

The disruption of this film causes the reinforcing steel to come in direct exposure to oxygen and moisture, in which corrosion will start to take place. During this stage, pitting of the reinforcement surface is common and is followed by final corrosion of the reinforcing steel. The formation of rust can cause swelling in the reinforcement and result in the cracking of the concrete cover.

However, the effect on concrete caused by the process of carbonation is observed to be slow, and is determined by the rate of carbon dioxide diffusing into the concrete. Although the rate of diffusivity depends highly on the water−cement ratio, level of hydration, humidity, temperature, permeability and porosity of the concrete, sufficient thickness of concrete cover over the steel reinforcement can prolong the life span of concrete pipes as it requires more time to penetrate through a thicker concrete, compared to a thinner concrete. With a thicker concrete layer, the reinforcement is further protected against corrosion, for longer periods of time.

Chloride Attack

In concrete pipes, chloride attack is widely recognized as the ability of chloride ions to enter into the concrete and induce corrosion of the steel reinforcement. It is by this action that, concrete pipes that are exposed to seawater often undergo corrosion faster than concrete pipes that do not have a simultaneous salt and water exposure. This creates a continuous corrosion favoring environment and with the oxide film lost in the previous carbonation process, due to a reduced pH, the rate of corrosion in the reinforcing steel is further heightened. An increased likelihood of chloride attack, however, is not only subject to seawater exposure. In fact corrosion of the reinforcing steel in concrete pipes is initiated when chloride concentrations exceed 0.6% of the binder mass of the concrete (Hajkova, 2015). A threshold concentration of 0.026% (by weight of concrete) is adequate enough to disrupt the oxide film on the surface of the steel reinforcement, allowing corrosion to take place (Daily, 2017).

Galvanic Corrosion

The process of galvanic corrosion in concrete pipes occurs once the oxide film on the surface of the steel reinforcement is disrupted. As the chloride ions do not react with the concrete surface in a uniform distribution, it creates a gradient effect, with some areas of the reinforcing steel exposed to a higher concentration of chlorides and some areas not exposed at all. The areas exposed to higher concentrations have an increased likelihood of corrosion initiation, causing a reduction in the oxide film, while the oxide film on reinforcing steel that is not exposed to chlorides remains intact.

Variances in the structural composition of the reinforcing steel, as well as variances in residual stress, can also initiate the process of galvanic corrosion. In this case, chlorides, reacting with the reinforcing steel in a uniform distribution, result in micro-cell corrosion, which is often localized and dominates the corrosion process. Under this type of action, the anodic and cathodic sites on the reinforcing steel are in close proximity to each other, therefore acting as a direct pathway for the constant continuation of the corrosion mechanism. This constant cycle results in loss of reinforcing steel material from anodic sites and creates pitting on the surface. As corrosion progresses the exposed pitted area of the reinforcing steel becomes less alkaline, which results in further loss of material from the bottom of the pitting. This loss of material is a "progressive" action and reaches a point at which the reinforcing steel is wasted away and no longer able to withstand applied loads, ultimately resulting in failure (Daily, 2017).

Corrosion Geometry

Uniform Corrosion Uniform corrosion is the most abundant in steel reinforcements, with the entire or large fraction of the surface area of the exposed steel bar evenly covered with rust, often observed as a reddish color. This reddish color can also be observed in concrete that is porous and wet, in which corrosion migrates through the concrete cover, and is commonly presented in the form of stains on the concrete surface. For dry concretes, however, corrosion can progress further by deforming the concrete cover and can result in spalling and cracking.

Pitting Corrosion

Pitting corrosion is characterized by a local formation of pin-sized holes over a small area on the surface of the reinforcement (Fig. 1.5). Due to the unpredictable nature of this type of corrosion, it is known to be very dangerous and normally exceeds the damage caused by uniform corrosion. The reason pitting corrosion is very common and unpreventable in steel reinforcements is due to difficulty in designing pipes to combat this effect. Therefore this type of corrosion often goes undetected as rust covers the pitting, rendering it unnoticeable.

In conditions that increase the likelihood of corrosion, the reinforcing steel in concrete pipes undergoes an initial corrosion reaction where it forms a passive oxide film on its surface. This film acts as a barrier between the steel and the

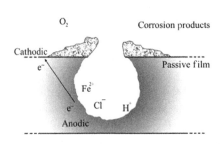
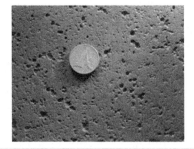

Figure 1.5 Pitting corrosion on steel element.

environment, slowing down the rate of further corrosion and thereby providing continued protection against corrosion.

Therefore, it is known that reinforcing steel normally does not corrode providing the oxide film is still intact. Factors that can alter the stability of the oxide film include pH. In freshly placed concrete, the hydration of cement enables the concrete to develop a high alkanity, and with oxygen reacts to stabilize the oxide film on the surface of the reinforcing steel. Due to the presence of calcium, potassium, and sodium hydroxides in concrete, a pH of above 12 is generally observed, which maintains the alkalinity and ensures the oxide film is preserved. However, carbonation and chloride attack are the two factors that cause corrosion in steel reinforcements of concrete pipes.

1.6.1.2 Microbial-Induced Corrosion (MIC)

Sewer pipes deteriorate at different rates depending on specific local conditions and it is not determined by age alone. There have been numerous cases of severe damage to concrete pipes, where it has been necessary to replace the pipes before the desired service life has been reached. There are many cases in which sewer pipes designed to last $50-100$ years have failed due to H_2S corrosion in $10-20$ years. In extreme cases, concrete pipes have collapsed in as few as 3 years (Pomeroy, 1976).

The deterioration of concrete itself consists of an entirely different deterioration mechanism. The concrete material in concrete pipes, does not simply just "corrode," as is the case with pipes involving metal.

The most corrosive agent that leads to the rapid deterioration of concrete pipelines in sewers is H_2S. Approximately 40% of the damage in concrete sewers can be attributed to biogenous sulfuric acid attack. Sulfide corrosion, which is often called microbiologically induced corrosion, has two distinct phases as follows:

- The conversion of sulfate in wastewater to sulfide, some of which is released as gaseous hydrogen sulfide.
- The conversion of hydrogen sulfide to sulfuric acid, which subsequently attacks susceptible pipeline materials.

The surface pH of new concrete pipe is generally between 11 and 13. Cement contains calcium hydroxide, which neutralizes the acids and inhibits formation of oxidizing bacteria when the concrete is new. However, as the pipe ages, the neutralizing capacity is consumed, the surface pH drops, and the sulfuric acid-producing bacteria become dominant. In active corrosion areas, the surface pH can drop to 1 or even lower and can cause a very strong acid attack. The corrosion rate of the sewer pipe wall is determined by the rate of sulfuric acid generation and the properties of the cementitious materials. As sulfides are formed and sulfuric acid is produced, hydration products in the hardened concrete paste (calcium silicon, calcium carbonate, and calcium hydroxide) are converted to calcium sulfate, more commonly known by its mineral name, gypsum (ASCE 69, 1989). The chemical reactions involved in sulfide build-up can be explained as follows.

Sulfate, generally abundant in wastewater, is usually the common sulfur source, though other forms of sulfur, such as organic sulfur from animal wastes, can also be reduced to sulfide. The reduction of sulfate in the presence of waste organic matter in a wastewater collection system can be described as follows:

$$SO_4^{-2} + ORGANIC\ MATTER + H_2O \underset{Bacteria}{\longrightarrow} 2HCO_3 - + H_2S$$

The H_2S gas in the atmosphere can be oxidized on the moist pipe surfaces above the water line by bacteria (*Thiobacillus*), producing sulfuric acid according to the following reaction (Meyer, 1980):

$$H_2S + O_2 \underset{Bacteria}{\longrightarrow} H_2SO_4$$

As sulfides are formed and sulfuric acid is produced, hydration products in the hardened concrete paste (calcium silicate, calcium carbonate, and calcium hydroxide) are converted to calcium sulfate. The chemical reactions involved in corrosion of concrete are

$$H_2SO_4 + CaO.SiO_2.2H_2O \rightarrow CaSO_4 + Si(OH)_4 + H_2O$$
$$H_2SO_4 + CaCO_3 \rightarrow CaSO_4 + H_2CO_3$$
$$H_2SO_4 + Ca(OH)_2 \rightarrow CaSO_4 + 2H_2O$$

Gypsum does not provide much structural support, especially when wet. It is usually present as a pasty white mass on concrete surfaces above the water line. As the gypsum material is eroded, the concrete loses its binder and begins to spall, exposing new surfaces. This process will continue until the pipeline fails or corrective actions are taken. Sufficient moisture must be present for the sulfuric acid-producing bacteria to survive, however; if it is too dry, the bacteria will become desiccated, and corrosion will be less likely to occur. Fig. 1.6 shows the process of sulfide build-up in a sewer system.

Although increasing the thickness of concrete pipes as well as the use of protective coatings can delay corrosion to an extent, it still remains inevitable that corrosion will occur and that failure is only postponed. Sewers also contain solvents and industrial wastes which can alter the pH and conditions in the concrete pipe significantly, causing deterioration, including crack formation and softening of the concrete.

Concrete Corrosion Rate
The rate of corrosion of a concrete sewer can be calculated from the rate of production of sulfuric acid on the pipe wall, which is in turn dependent upon the rate

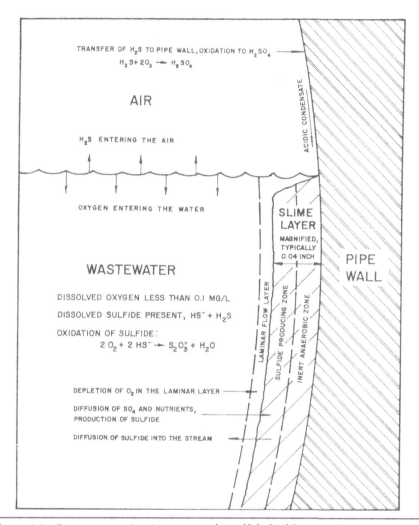

Figure 1.6 Process occurring in sewer under sulfide build up conditions (ASCE No. 69, 1989).

that H_2S is released from the surface of the sewage stream. The average flux of H_2S to the exposed pipe wall is equal to the flux from the stream into the air multiplied by the ratio of the surface area of the stream to the area of the exposed pipe wall, which is the same as the ratio of the width of the stream surface (b) to the perimeter of the exposed wall ($\acute{P}$). The average flux of H_2S to the wall is therefore calculated as follows (Pomeroy, 1976):

$$\Phi = 0.7(su)^{3/8} j[DS](b/\acute{P}) \tag{1.17}$$

where s is pipe slope, u is velocity of stream (m/s), j is pH-dependent factor for proportion of H_2S, [DS] is dissolved sulfide concentration (mg/L). A concrete pipe is made of cement-bonded material, or acid-susceptible substance, so the acid will penetrate the wall at a rate inversely proportional to the acid-consuming capability (A) of the wall material. The acid may partly or entirely react. The proportion of acid that reacts is variable (k), ranging from 100% when the acid formation is slow, to perhaps 30%−40% when it is formed rapidly. Thus, the average rate of corrosion (mm/year) can be calculated as follows

$$c = 11.5k\Phi(1/A) \qquad (1.18)$$

where k is the factor representing the proportion of acid reacting, to be given a value selected by the judgement of the engineer and A is the acid-consuming capability, alkalinity, of the pipe material, expressed as the proportion of equivalent calcium carbonate. The A value for granitic aggregate concrete ranges from 0.17 to 0.24 and for calcareous aggregate concrete, A ranges from 0.9 to 1.1 (ASCE No.60, 2007). Substituting Eq. (1.17) into Eq. (1.18):

$$c = 8.05k \times (su)^{3/8}j \cdot [DS] \times \frac{b}{P'A} \qquad (1.19)$$

Therefore the reduction in wall thickness in elapsed time t, is:

$$d(t) = c \cdot t = 8.05k \cdot (su)^{3/8}j \cdot [DS] \times \frac{b}{P'A} \cdot t \qquad (1.20)$$

1.6.2 DETERIORATION OF FERROUS PIPES

Ferrous pipes, including cast iron, wrought iron, ductile iron, and mild steel, all hold a similarity in the sense that they are all derived from metals. Each of these pipes hold varying structural and physical properties, and show differences in strength.

However, the different types of ferrous pipes all show the same deterioration mechanism, even though corrosion rates may differ depending on the various conditions these materials present. The various conditions include the presence of factors that favor corrosion, or pipes holding particular structural characteristics that are not able to withstand high levels of stresses and can readily lead to failure.

Therefore, the degree of corrosion occurring in these different types of ferrous pipes and the severity of the damage caused as a result would depend on where these pipes are used, the environments they are subject to as well as differences in the structural composition, (i.e., carbon content) and design methods. For example, the corrosion rate in ferrous pipes carrying freshwater compared to wastewater can differ, due to a number of given various conditions, including

ion^+ content, pH, conductivity, temperature variations, and humidity. Similarly, the level of corrosion occurring in ferrous pipes in petrochemical plants would be very different to ferrous pipes that carry only "mild" corrosive chemicals.

The level of corrosion resulting in failure of these pipes would also depend on the factors stated above, for example, the failure rate of ductile iron may be different than wrought iron. If the failure rate was conducted only on the basis of strength, given that wrought iron is pure and therefore may last longer, it may take longer than ductile iron to fail. However, under various conditions to which these two types of pipes are subject, such as corrosive environments, failure rates may be prone to change.

It is known that protective coatings and linings are used in ferrous pipes to protect them against caustic environments, such as exposure to aggressive soils, and corrosive agents. Without the presence of these protective layers covering the surface of these metals, it allows a direct path for corrosion to occur. Also various factors including temperature fluctuations, humidity, ion^+ content, pH, and exposure to corrosive substances can break down the protective coatings and linings.

In water distribution systems, although cast iron pipes are being phased out of the water pipeline networks in many countries, a significant portion of current networks are comprised of cast iron pipes with some of them up to 150 years old. There are approximately 335,000 km of water mains in the United Kingdom and more than 60% is estimated to be cast iron pipes (Water UK, 2007). In the United Kingdom, the failure rate of cast iron pipes can be as high as 3000 failures per year (i.e., 10 bursts/1000 km/year) (UKWIR, 2002). Of many mechanisms for pipe failures, corrosion of cast iron has been found to be the most predominant, which is linked to almost all pipe failures (Misiunas, 2005).

Global data shows that, on average, cast iron has been the dominating material for water distribution pipes before the 1960s. Therefore the average age of cast iron pipes in existing networks has been estimated to be 50 years (Rajani and Kleiner, 2004; Misiunas, 2005). Due to their long-term use, the aging and deterioration of pipes are inevitable and indeed many failures have been reported worldwide (Atkinson et al., 2002; Misiunas, 2005; Rajani and Tesfamariam, 2007; EPA/600, 2012). Depending on the country, compared with other types of pipe material, cast iron pipes have the highest frequency of breaks as shown in Table 1.4. It has been established (e.g., Yamini and Lence, 2009; EPA/600, 2012) that the corrosion of cast iron is the most common form of deterioration of the pipes and it is a matter of concern for both the safety and serviceability of pipes.

The predominant deterioration mechanism of iron-based pipes is electrochemical corrosion with the damage occurring in the form of corrosion pits. The damage to iron is often identified by the presence of graphitization, a result of iron being leached away by corrosion. Either form of metal loss represents a corrosion

TABLE 1.4 Frequency of Pipe Breakage for Different Materials
(Breaks/100 km/year) (Misiunas, 2005)

Source	Cast Iron	Ductile Iron	PVC
NRC (1995)	36	9.5	0.7
Weimer (2001)	27	3	4
Pelletier et al. (2003)	55	20	2

pit that grows with time and reduces the thickness and mechanical resistance of the pipe wall. This process eventually leads to the breakage of the pipe.

Certain factors that can act as stimulators to the corrosion of buried ferrous pipes include moisture content, temperature, and pH levels, as well as the presence of sulfides and mineral salts. The rate at which corrosion occurs depends highly on the physical and chemical composition of soils, however, corrosion of metal pipelines in soil environments occurs as an effect of the combined factors mentioned above.

In fact, the minerals in naturally occurring soils can accelerate the corrosion process in pipes that are exposed to water and air by forming salt, which is a key component that allows corrosion to occur. This mechanism involves the formation of rust, which produces salt on the exposed area of the pipe and causes this action to further feed into the corrosion process. The process is especially heightened in salt-rich soils, which are shown to be more acidic and can create a more corrosive environment.

On the subject of corrosion, it is important to understand how the process links to failure mechanisms. Pipes derived from metal consist of two main failure modes, including rupture caused by wall thickness reduction and fracture caused by stresses concentrated on the tips of cracks that are already present. Corrosion is an example that produces these cracks. The two main failure modes correspond to the effects of corrosion by affecting the fracture toughness and tensile strength of pipes. In fact it is observed that most pipes affected by corrosion show failure caused by fracture rather than loss of strength.

The actual failure caused by fracture is due to the growth of the crack, which continues until the pipe collapses. Therefore, understanding both fracture toughness and tensile strength of metals will enable a better prediction of pipe failure to be drawn (Hou, et al, 2016).

Before discussing various factors that contribute to the corrosion process and that accelerate this mechanism, the corrosion mechanism itself needs to be fully examined.

Corrosion causes a change in the composition of a metal and creates a less desirable material which can lead to a loss of function either in the exposed section of the metal or the system as a whole. In ferrous pipes, the process of corrosion is seen as rust which forms on the surface of the metal. The process starts when atoms in the metal undergo changes including loss of electrons and become positively charged. This positive charge of atoms attracts atoms of negative charge, creating bonds, leading to the inevitable start of corrosion, providing water and oxygen are present to allow the process to follow through. New material begins to form on the surface of the metal and continues to react if conditions that feed corrosion are continued.

Iron (Fe) is the major component. The formula below shows the negatively charged material reacting with iron and surrounding electrons.

$$O_2 + 2H_2O + 4e^- \rightarrow 4OH^-$$

This reaction gets used up to produce the formula below, which shows the reaction between iron, water, and oxygen to produce iron hydroxide.

$$2Fe + O_2 + 2H_2O \rightarrow 2Fe(OH)_2$$

It can be observed by the second equation, that as oxygen readily dissolves in water, it further reacts with the iron hydroxide to produce hydrated iron hydroxide (rust).

$$4Fe(OH_2) + O_2 \rightarrow 2H_2O + 2Fe_2O_3 \cdot H_2O$$

The main reason why corrosion occurs in pipes that are exposed to water is due to the corrosion process needing water which creates a path for the movement of ions and electrons. This process is further heightened by conditions that give rise for these reactions to occur, including factors that act as driving forces such as applied stresses. Depending on conditions, stresses can account for localized corrosion which is the cause of 70% of failures and uniform corrosion causing 30% of failures (Li and Ni, 2013). Systems with localized corrosion can have greater consequences than those with uniform corrosion as failures are spontaneous and do not follow the "overuse"-caused failure phenomenon. It can occur within a short period of usage which makes localized corrosion a main target to pipelines that are subject to corrosive environments.

Factors such as changes in fluid flow, turbulence, spontaneous directional changes, and materials which obstruct flow in pipes can cause a corrosive environment and accelerate corrosion in pipelines. As mentioned previously, the addition of coatings and linings can prevent corrosion to an extent and prolong the lifetime of pipes, however, fluid flow and added pressures can wash away the protective layers and expose the underlying metal pipes, and highlight the risk of corrosion.

Another factor that contribute to corrosion is fatigue, caused simply by cyclic stresses due to sudden changes in fluid flow and pressures in pipes. However, a combination of factors, including cyclic stresses and a favorable environment for corrosion, is more damaging than fatigue alone, which is usually the case in pipelines. The usage of protective coatings and designing pipes with a focus on reducing stresses that concentrate on particular areas, avoiding spontaneous changes and removing potential causes of cyclic stresses can reduce the risk of corrosion. However, in pipelines where fatigue is an issue, failure is inevitable.

Another inevitable case of failure is stress corrosion cracking (SCC), which consists of many functions that cause and accelerate this type of cracking. Pipes that are subject to corrosive environments are prone to the formation of cracks on the surface, which grow and cause spontaneous failure. Although metal pipes are ductile and undergo tensile stress on a regular basis, SCC occurs, particularly when combined with elevated temperatures. This occurrence is more common in pipes made from alloys, such as galvanized iron, than pure metals. It is observed that temperature fluctuations can alter the temperature gradient and accelerate the rate of corrosion. This is very common in pipes that carry fluids of varying temperatures and hot water pipes, caused by thermogalvanic corrosion. Vapor caused by the evaporative action in pipes with temperature variances create an almost constant moist environment and a thriving atmosphere for corrosion.

Stress corrosion cracking is also a known problem for pressure pipes derived from stainless steel, especially common in chemical and petrochemical plants. The presence of aggressive agents favors the corrosive environment and stainless steel is particularly susceptible to corrosion when temperatures are exceedingly high. Although stainless steel contains carbon content of 0.03 wt%, which is significantly lower than most metal alloys, along with a high chromium content, which allows resistivity to corrosion, this effect is counteracted when exposure to high temperatures of 415−850°C. This causes a change in the microstructure of stainless steel which becomes susceptible to precipitation of chromium carbides, causing sensitization. The formation of these carbides cause a depletion in the chromium content of sensitized steels, resulting in a rapid change in its structural composition, allowing an increased susceptibility to failure.

However, for stainless steels, environments that cause cracking are specific as not all environments cause SCC. The ability of stainless steels to resist corrosion is due to the formation of a passive layer which occurs in oxidizing environments. This occurs when the steel consists of a 10.5% minimum chromium content. However, corrosion affecting stainless steels is different to corrosion of other metals. For corrosion to occur in stainless steels a change or break down of the passive layer in environments that present nonoxidizing conditions is required. Environments containing caustic chemicals, changes in pH, and temperature, as well as contamination, fabrication, and maintenance processes can affect the passive layer and the corrosion rate, as well as the type of corrosion that takes place in stainless steels.

The passive layer in stainless steel is susceptible to attacks from chlorides, caustic acid, and polythionic acid, which generally occur at localized points. Stainless steels consisting of a higher level of alloy material, such as the combination of nickel, have a higher susceptibility to corrosion in chloride-rich environments. According to a study conducted by Khalifeh et al. (2017), a high incidence of failures was observed in equipment derived from these materials.

However, for SCC to occur, the presence of tensile stresses is required, which can be either applied, thermal, or residual and it is usually the combined effect that accelerates SCC.

For SCC to take place solely as a result of applied stresses, a very high magnitude of these stresses is needed. On the contrary the effect of residual stresses was observed to cause a higher susceptibility to SCC.

Residual stresses are defined by the tensile or compressive force of a material in the absence of any other external loading. The process of welding and mechanical forces in the production of these materials can cause residual stress. The expansion and contraction action caused by temperature variations that occur in welding are what cause residual stress. The mechanical forces which occur during the fabrication process of these materials cause residual stress by the bending action of steel during pipe formation. This process can also cause localized deformations, which can lead to an early onset of SCC.

In sensitized stainless steels, the formation of chromium carbides also causes overheating, and high temperatures are known to feed the growth rate of crack formation. As a result, these cracks further allow chlorides and other caustic agents to leak straight through and allow SCC.

1.6.2.1 Corrosion Rate in Ferrous Pipes

Corrosion pits have a variety of shapes with characteristic depths, diameters (or widths), and lengths. They can develop randomly along any segment of pipe and tend to grow with time at a rate that depends on environmental conditions in the immediate vicinity of the pipeline (Rajani and Makar, 2000).

The corrosion rate of in-service ferrous pipes is believed to be higher in the beginning and then decreases over time as corrosion appears to be a self-inhibiting process (Shreir et al., 1994). Furthermore, due to the variation of service environment it is rare that the corrosion occurs uniformly along the pipe but is more likely locally in the form of a corrosion pit.

A number of models for corrosion of ferrous pipes have been proposed to estimate the depth of corrosion pit (e.g., Sheikh et al., 1990; Ahammed and Melchers, 1997; Kucera and Mattsson, 1987; Rajani et al., 2000; Sadiq et al., 2004). For example, Sheikh et al. (1990) suggested a linear model for corrosion growth in predicting the strength of cast iron pipes. A decade later, Rajani et al.

(2000) proposed a two-phase corrosion model where the first phase is a rapid exponential pit growth and the second is a slow linear growth. There are debates in the research community as to whether the corrosion rate can be assumed linear or otherwise. A widely accepted model of corrosion as measured by the depth of corrosion pit is of a power law which was first postulated for atmospheric corrosion by Kucera and Mattsson (1987) and can be expressed as follows:

$$a = kt^n \tag{1.21}$$

where t is exposure time and k and n are empirical constants largely determined from experiments and/or field data.

For underground corrosion, the constants are typically functions of localized conditions, including soil type, the availability of oxygen and moisture, and properties of pipeline material. In many cases it may be possible to use past experience to derive estimates for the two constants in Eq. (1.21), but with somewhat more effort than would be necessary to estimate a constant corrosion rate as used conventionally (Ahammed and Melchers, 1997).

Rajani et al. (2000) proposed a two-phase corrosion model to accommodate this self-inhibiting process:

$$A = \alpha t + \beta(1 - e^{-\lambda t}) \tag{1.22}$$

where α, β, and λ are constant parameters.

In the first phase of the above equation there is a rapid exponential pit growth and in the second phase there is a slow linear growth. This model was developed based on a data set that lacked sufficient points in the early exposure times. Therefore prediction of pit depth in the first $15-20$ years of pipe life should be considered highly uncertain when Eq. (1.22) is used.

An example of field data which shows the rate of internal and external corrosion for cast iron pipes has been illustrated in Fig. 1.7 (Marshall, 2001). As it can be concluded from this data, external corrosion has a higher rate than internal corrosion, especially during early stages. In Fig. 1.8, a sample of a cast iron pipe taken from London water mains from the Victorian era (i.e., $1800-1900$) also shows the severity of external corrosion compared with internal corrosion.

1.6.3 DETERIORATION OF PLASTIC PIPES

Due to metals holding a higher susceptibility to corrosion, plastic piping can be particularly useful in environments that are corrosive, including for pipes that experience a high occurrence of temperature fluctuations.

These fluctuations create an "evaporative" action, causing vapor to form inside pipes. This results in constant moisture inside the pipe, providing

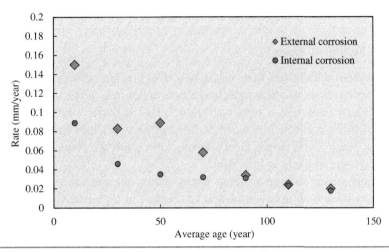

Figure 1.7 Rate of internal and external corrosion for cast iron pipes, Marshall (2001).

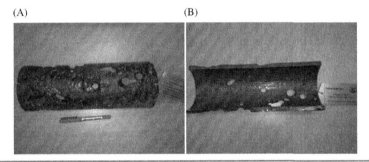

Figure 1.8 A section of London's Victorian water mains: (A) external corrosion; (B) internal corrosion.

opportunity for the corrosion to occur. However, this may not be the case with plastic pipes, including PVC, CPVC, GRE, and HDPE. Although plastic pipes may have a lower strength compared to pipes derived from metals, they provide the combined benefit of strength that is required for the proper function of these pipes, as well as reduced susceptibility to corrosion.

Despite of resistance to corrosion, the lower strength of these types of pipes enable deflections and deformations, which can cause cracking and failure. Because plastic pipes do not interact with water and air in the same way as pipes derived from metal, plastic pipes are particularly useful in cold water systems.

Exposure to harsh chemical environments, such as acetones, ketones, and strong hydrocarbons, can cause the walls of the pipe to dissolve or thin out over time, if prolonged exposure were to occur.

1.6.4 DETERIORATION OF OTHER TYPES OF PIPES

1.6.4.1 Seamless Pipes

Seamless pipes are derived from metal, however, they are also available with a plastic coating inside the steel pipe. In the case of seamless pipes the presence of plastic linings inside these pipes can allow pipes to have a relatively higher strength than most plastic pipes, and at the same time maintain the resistance to corrosion. There is also potential for corrosion of metal if the plastic cracks, causing the exposure of metal to fluid flow.

Seamless pipes undergo extreme forces during the manufacturing process; therefore, a thick high-pressure lubricant is applied to the surface of the pipe to prevent cracking, which can result in a defective pipe. Following this process, the pipe undergoes heat treatment in which the high pressure lubricant must be removed. The aggressive chemicals used to dissolve the solvent on the surface of the pipe can still remain as a residue long after the manufacturing process is completed and can result in corrosion. This is especially problematic in seamless pipes with thin diameter walls.

As tensile and yield strengths are related to the thickness of the pipe wall thickness, it is important that this thickness is not depleted in the slightest due to corrosion, otherwise it can result in pipe failure.

Wall thickness is also important in terms of heat transfer, as these types of pipes can be prone to temperature fluctuations, as well as high heats; this can be particularly problematic if manufacturing processes are not properly followed.

1.6.4.2 Glass Reinforced Epoxy Resin Composite (GRE)

GRE used in pipelines are known to be corrosion resistant and the cause of failure is often the result of poor design methods. Although using this material for pipe production is effective and capable of withstanding most waters and alkali environments, the issue lies within exposure to strong acids. This exposure alone however may not be enough to cause deterioration, but rather becomes possible with a combined effect of poor design methods and applied loads.

With GRE, the choice of glass fiber is critical. Given an unsuitable glass fiber type is used, exposure to corrosive chemicals can cause the fiber to break down and break the resin, altering the structural composition of the composite. This can result in a structurally inadequate material and if used in pipes, can alter their strength and ability to resist corrosion.

The deterioration rate of the glass fibers exposed to corrosive environments depends on the level of diffusion and osmosis of corrosive chemicals into the fibers. Various factors, including thermal gradients and pressure gradients, can change this effect, however the resulting deterioration may occur in a longer time-frame, in comparison to effects observed with a combined action of these gradients and applied stresses, which can enhance or reduce the level of diffusion and osmosis and thereby the resulting degree of deterioration.

Applied stresses alone can also contribute to deterioration of GRE, given that the capacity to withstand the applied stresses is exceeded, result in swelling, embrittlement, and microcracking of the structure. The continuous impact of these factors over a long period of time can cause exhaustion of the structure, and increase the susceptibility towards various failure modes, ultimately resulting in deterioration (Owens Corning, 2011).

1.6.4.3 Clay Pipes

The extent of resistivity to the damaging effects of sulfuric acid in piping highly depends on the pipe material. Pipes derived from cement and metals hold a higher susceptibility to rapid deterioration when exposed to sulfuric acid. This is especially common in sewer systems, where concrete piping is typically used and is subject to corrosion due to the highly acidic nature of the environment. The growth of bacteria and microbes in these pipes, which also generate sulfur, is a component that adds to the sulfur-based corrosion that occurs inside these concrete pipes. Factors including acidic soils and temperature changes also accelerate this process and provide a thriving environment to conditions that favor corrosion.

The usage of vitrified clay pipes in sewer systems have proven to be the only pipe material to resist sulfur-based corrosion, as well as aggressive soils, solvents, and industrial wastes. Due to this resistivity, no protective coating and linings are necessary.

Vitrified clay pipes are also known for their rigidity and ability to avoid loss of strength by not deflecting or flattening when loads are applied, in comparison to plastic pipes which deflect and deform under loads (Gladding McBean, 2017).

References

American Concrete Pipe Association, ACPA, 1993. Concrete pipe technology handbook, Irving, Texas.

Ahammed, M., Melchers, R.E., 1994. Reliability of underground pipelines subject to corrosion. J. Transport. Eng. 120 (6), Nov/Dec.

Ahammed, M., Melchers, R.E., 1997. Probabilistic analysis of underground pipelines subject to combined stress and corrosion. Eng. Struct. 19 (12), 988–994.

Alton Materials, 2017. The difference between ferrous and Non Ferrous scrap metal. [online] Available from: http://www.altonmaterials.com/the-differences-between-ferrous-and-non-ferrous-scrap-metal/ (Accessed 25 April).

Anon, 2002. Office of Pipeline Safety, websites: www.ntsb.gov and www.ops.dot.gov

ASCE 15-98, 2000. Standard practice for direct design of buried precast concrete pipe using standard installations (SIDD), Reston, VA.

ASCE Manuals and Reports of Engineering Practice - No. 69, 1989. Sulphide in Wastewater Collection and Treatment Systems. American Society of Civil Engineers.

ASCE Manuals and Reports of Engineering Practice No. 60, 2007. Gravity Sanitary Sewers, second ed. American Society of Civil Engineers, New York, USA.

ASTM E632-82, 1996. Standard Practice for Developing Accelerated Tests to Aid Prediction of the Service Life of Building Components and Materials.

Asi, O., 2009. Mechanical properties of glass-fiber reinforced epoxy composites filled with Al2O3 particles. J. Reinforced Plastics Composites 28 (23), 2861−2867.

Atkinson, K., Whiter, J.T., Mulheron, M.J., Smith, P.A., 2002. Failure of small diameter cast iron pipes, Urban. Water 4, 263−271.

BS EN 1295-1, 1997. Structural design of buried pipelines under various conditions of loading, Part 1: general requirements, BSI.

Daily, S.F., 2017. Understanding corrosion and cathodic protection of reinforced concrete structures. [Online] Available from: http://beta.corrpro.com/-/media/Corporate/Files/Corrpro-Literature/Technical-Papers/Understanding-Corrosion-and-Cathodic-Protection-of-Reinforced-Concrete-Structures.ashx [Accessed 5 May].

Davis, P., De Silva, D., Gould, S., Burn, S., 2005. Condition assessment and failure prediction for asbestos cement sewer mains, Pipes Wagga Wagga Conference, Charles Sturt University, Wagga Wagga, New South Wales, Australia.

De Belie, N., Monteny, J., Beeldens, J.A., Vincke, E., Van Gemert, D., Verstraete, W., 2004. Experimental research and prediction of the effect of chemical and biogenic sulfuric acid on different types of commercially produced concrete sewer pipes. J. Cement Concrete Res. 34, 2223−2236.

De Silva, D., Moglia, M., Davis, P., Burn, S., 2006. Condition assessment and probabilistic analysis to estimate failure rates in buried metallic pipelines. J. Water Supply: Res. Technol.-Aqua 55 (3), 179−191.

EPA/600/R-12/017, 2012. Condition assessment technologies for water transmission and distribution systems, United States, Environmental Protection Agency.

Ferrostaal, 2017. Seamless pipes and tubes (SMLS). [Online] Available from: http://www.ferrostaal-piping.com/en/products-bu-piping-ferrostaal/industrial-pipe-bu-piping-ferrostaal/seamless-pipes-bu-piping-ferrostaal/ [Accessed 5 May 2017].

Gladding McBean, 2017. Why clay pipe? [Online] Available from: https://www.gladdingmcbean.com/sewer-pipe/why-clay-pipe [Accessed 5 May].

Grigg, N.S., 2003. Infrastructure management systems. Water, Wastewater, and Stormwater Infrastructure Management. Lewis Publishers, Boca Raton, FL, pp. 1−17.

Hajkova, K., 2015. Mechanism of reinforcement. [Online] Available from: http://ksm.fsv.cvut.cz/~nemecek/teaching/dmpo/clanky/2015/Hajkova_Karolina.pdf [Accessed 5 May 2017].

Hou, Y., Lei, D., Li, S., Yang, W., Li, C.-Q., 2016. Experimental investigation on corrosion effect on mechncial properties of burried metal pipes. Int. J. Corrosion 2016 (2016), 1−13.

JSTEEL, 2017. Seamless pipe. [Online] Available from: http://www.jsteel.com.au/products/tubular-piles-and-pipes/seamless-pipe/ [Accessed 4 May 2017].

Khalifeh, A.R., Dehghan Banaraki, A., Daneshmanesh, H., Paydar, M.H., 2017. Stress corrosion cracking of a circulation water heater tubesheet. Eng. Failure Analysis 78, 55−66.

Kucera, V., Mattsson, E., 1987. In: Mansfeld, F. (Ed.), Atmospheric Corrosion, in Corrosion Mechanics. Marcel Dekker Inc, New York.

Kumar, S., Panda, A.K., Singh, R.K., 2011. A review on tertiary recycling of high-density polyethylene to fuel. Resourc. Conserv. Recycling 55 (11), 893−910.

Li, C., Ni, Y., 2013. An integrated probabilistic approach for prediction of remaining life of buried pipes, in Drahor Noak & Miroslav Vorechovsky (Ed.), Proceedings of the 11th International Probabilistic Workshop, Brno, Czech Republic, 6−8 November 2013, pp. 41−52.

Mahmoodian, M., Alani, A., 2014. Modelling deterioration in concrete pipes as a stochastic Gamma process for time dependent reliability analysis. ASCE J. Pipeline Systems Eng. Practice 5 (1), 04013008.

Mahmoodian, M., Alani, A., 2016. Sensitivity analysis for failure assessment of concrete pipes subjected to sulphide corrosion. Urban Water J. 13 (6), 637−643.

Mahmoodian, M., Li, C.Q., 2017. Failure assessment and safe life prediction of corroded oil and gas pipelines. J. Petrol. Sci. Eng. 151 (2017), 434−438.

Marshall, P., 2001. The Residual Structural Properties of Cast Iron Pipes - Structural and Design Criteria for Linings for Water Mains. UK Water Industry Research.

Meyer, W.J., 1980. Case study of prediction of sulphide generation and corrosion in sewers. J. Water Pollution Control Federation 52 (11), 2666−2674.

Misiunas, D., 2005. Failure Monitoring and Asset Condition Assessment in Water Supply Systems, PhD dissertation. Department of Industrial Electrical Engineering and Automation Lund University.

Moglia, M., Davis, P., Burn, S., 2008. Strong exploration of a cast iron pipe failure model. J. Reliabil. Eng. System Safety 93, 863−874.

Moser, A.P., Folkman, S., 2008. Buried Piep Design, third ed. Mc Graw Hill, Europe.

NRC, 1995. Water Mains Break Data on Different Pipe Materials for 1992 and 1993, Technical Report A-7019.1, National Research Council Canada.

OFWAT, 2002. Maintaining Water and Sewerage Systems in England and Wales, Our Proposed Approach for the 2004 Periodic Review, London.

Owens Corning, 2011. Glass fibre reinforcement resistance guide. [Online] Available from: http://www.owenscorningindia.com/ocindia/composites/pdf/oc_chemical_resistance_guide_edition.pdf [Accessed 5 May].

Pearlite Steel, 2017. Difference Between Seamless and ERW Stainless steel pipe. [Online] Available from: http://pearlitesteel.com/erw-stainless-steel-pipe-manufacturer-from-india/ [Accessed 4 May 2017].

Pelletier, G., Milhot, A., Villeneuve, J.P., 2003. Modelling water pipe breaks- three case studies. J. Water Resour. Planning Manage. 129 (2), 115−123.

Pomeroy, R.D., 1976. The Problem of Hydrogen Sulphide in Sewers. Clay Pipe Development Association.

Rajani, B., Kleiner, Y., 2004. Non-destructive inspection techniques to determine structural distress indicators in water mains, Evaluation and Control of Water Loss in Urban Water Networks, Valencia, Spain, June 21−25, 2004, pp. 1−20.

Rajani, B., Makar, J., 2000. A methodology to estimate remaining service life of grey cast iron water mains. Canadian J. Civil Eng. 27 (6), 1259−1272.

Rajani, B., Makar, J., McDonald, S., Zhan, C., Kuraoka, S., Jen, C.K., et al., 2000. Investigation of Grey Cast Iron Water Mains to Develop a Methodology for Estimating Service Life. American Water Works Association Research Foundation, Denever, CO.

Rajani, B., Tesfamariam, S., 2007. Estimating time to failure of Cast Iron water mains. Water Manage. 160 (WM2), 83−88.

Ridgers, D., Rolf, K., Stål, O., 2006. Management and planning solutions to lack of resistance to root penetration by modern PVC and concrete sewer pipes. Arboricult. J. 29 (4), 269−290.

Sadiq, R., Rajani, B., Kleiner, Y., 2004. Probabilistic risk analysis of corrosion associated failures in cast iron water mains. Reliab. Eng. System Safety 86 (1), 1−10.

Salman, B., Salem, O., 2012. Modeling Failure of Wastewater Collection Lines Using Various Section-Level Regression Models. ASCE J. Infrastruct. Syst. 18 (2), June 1.

Saltelli, A., Tarantola, S., Campolongo, F., Ratto, M., 2004. Sensitivity Analysis in Practice: A Guide to Assessing Scientific Models. Wiley, New York.

Sharma, K.J., Yuan, Z., de Haas, D., Hamilton, G., Corrie, S., Keller, J., 2008. Dynamics and dynamic modelling of H_2S production in sewer systems. Water Res. 42, 2527−2538.

Sheikh, A.K., Boah, J.K., Hansen, D.A., 1990. Statistical modelling of pittingcorrosion and pipeline reliability. Corrosion-NACE. 46 (3), 190−197.

Shreir, L.L., Jarman, R.A., Burstein, G.T., 1994. Corrosion. Butterworth Heinemann, London.

Sinha, S.K., Pandey, M.D., 2002. Probabilistic neural network for reliability assessment of oil and gas pipelines. Comput-Aided Civil Infrastruct. Eng. 17, 320−329.

The Urban Waste Water Treatment Directive, 91/271/EEC, 2012. Waste water treatment in the United Kingdom − 2012 Implementation of the European Union Urban Waste Water Treatment Directive − 91/271/EEC, Department for Environment, Food and Rural Affairs, UK.

The World Factbook, 2010. Central Intelligence Agency, USA, https://www.cia.gov/library/publications/the-world-factbook/fields/2117.html

Tran, D.T., 2014. Reliability-based Structural Design of Concrete Pipes. J. Fail. Anal. Prev. 14 (6), 818−825.

UKWIR, 2002. The residual structural properties of cast iron pipes structural and design criteria for linings for water mains, 01/WM/02/14.

Vincke, E., 2002. Biogenic Sulfuric Acid Corrosion of Concrete: Microbial Interaction, Simulation and Prevention. Ph.D. Thesis. Faculty of Bio-engineeringScience, University Ghent, Ghent, Belgium, pp. 7−9.

Water UK, 2007. Working on behalf of the Water Industry for a sustainable future, Water UK, http://www.water.org.uk/home/resources-and-links/waterfacts/resources.

Weimer, D., 2001. Water loss management and techniques, Technical report, DVGW, The German Technical and Scientific Association for gas and water.

Whidden, W.R., 2009. Buried Flexible Steel Pipe - Design and Structural Analysis. American Society of Civil Engineers (ASCE).

WIN, 2000. Clean and Safe Water for the 21st Century − A Renewed National Commitment to Water and Waste Water Infrastructure. Water Infrastructure Network, Washington, DC.

WSAA, Water Services Association of Australia, 2009. National Performance Report.

XINLIN, 2014. Differences between seamless and non-seamless pipes. [Online]. Available from: http://www.xinlinsteel.com/news/news_show-242.html [Accessed 5 May].

Yamini, H., Lence, B.J., 2009. Probability of failure analysis due to internal corrosion in cast-iron pipes. J. Infrastruct. Syst. 16 (1).

Zhang, L., De Schryver, P., De Gusseme, B., De Muynck, W., Boon, N., Verstraete, W., 2008. Chemical and biological technologies for hydrogen sulphide emission control in sewer systems: a review. Water Res. 42 (1-2), 1−12.

Zhou, W., 2011. Reliability evaluation of corroding pipelines considering multiple failure modes and time-dependent internal pressure. J. Infrastruct. Syst. 17 (4), 216−224.

Further Reading

HDPE, 2017. High-Density Polyethylene Pipe Systems. [Online] Available from: https://plasticpipe.org/pdf/high_density_polyethylene_pipe_systems.pdf [Accessed 26 April 2017].

Makar, J.M., Desnoyers, R., McDonald, S. E., 2001. Failure modes and mechanisms in gray cast iron pipe, NRCC-44218, Underground Infrastructure Research: Municipal, Industrial and Environmental Applications, Proceedings, Kitchener, Ontario, pp. 1−10.

Marston, A., Anderson, A.O., 1913. The Theory of Loads on Pipes in Ditches and Tests of Cement and Clay Drain Tile and Sewer Pipe. Bulletin 31. Iowa Engineering Experiment Station, Ames.

Pipeline Inspection and Maintenance

Chapter 2

Abstract

In this chapter, first, different failure modes which need to be considered in reliability and integrity assessment of pipelines will be discussed considering their causes. Common formulations for failure modes (limit state functions) will be presented for some types of pipes. This can give an idea to inspection engineers about the data which need to be monitored during their inspections for a more accurate reliability analysis.

Afterwards in Section 2.3, common methods for pipeline inspection will be introduced and the pros and cons of each technique and their application will be discussed and compared.

Lastly, pipeline maintenance methods to prevent pipelines from deterioration and to extend their service lives will be explained.

Chapter Outline

49

Reliability and Maintainability of In-Service Pipelines. DOI: https://doi.org/10.1016/B978-0-12-813578-5.00002-0

2.1 Background

Pipeline infrastructures are valuable assets and therefore the inspection and continuous maintenance of pipes must be actively carried out in order to guarantee the long service life and the best pipeline operation. Current literature in this field has understood that the need for an automatic defect detection technology is necessary since the use of human operators and manual inspection process is very time-consuming and can only capture 60%–75% of data accurately without the use of automated system in place. Although automated systems such as intelligent pigging have been developed, further work and innovation must be undertaken in order to create a complete automated inspection process that is able to be used in pipeline networks accurately and produce reliable data on a consistent basis.

The reliability of a pipeline can be assessed using the data collected through routine pipeline inspections. The reliability of a pipeline defines how well the pipeline is functioning and how well it will remain functioning at its very best without failure. Further explanation on pipeline reliability is found in Chapter 3, Methods for Structural Reliability Analysis.

Corrosion is a major factor which decreases pipeline reliability. Thus monitoring corrosion progress through inspection should be of more interest for pipeline maintenance engineers.

Inspection can be more challenging for underground pipelines, as they are hidden without easy access for inspection. Many of these assets are approaching their design lifetimes and pose a potential risk for structural failure due to defective problems such as corrosion and deterioration. The continuous monitoring, assessment, and evaluation of these pipelines is essential to ensure the networks are safe and have no serious problems.

Much of the current available literature acknowledge that manual inspection methods are extremely time-consuming and are also error-prone in detecting defects in the pipelines and therefore are not very reliable or accurate. However, developments in the inspection techniques are constantly growing and as a result, automated inspection and evaluation systems are replacing the manual inspection procedures to resolve this concern.

Pipeline inspection can be very expensive to carry out and it is crucial to follow reliable methods to maintain the pipes' safety and prevent defects from occurring in the future and reduce the total costs of pipeline maintenance. Common methods for maintenance of pipelines such as coatings and cathodic protection will be discussed in more detail in Section 2.4.

2.2 Pipelines Integrity

Pipeline operators and regulators need to carefully consider the integrity management of pipelines. Integrity management plans include emergency protocols but most importantly include maintenance and preventative measures. Operators take measures to defend pipelines against specific risks. Considering the potential for costly public and environmental damages given a pipeline failure, these measures are generally mandated by government. Through proper maintenance and inspection, operators can manage the integrity of their pipelines and mitigate the risk of failure.

Pipelines might lose their integrity often due to corrosion, manufacturing problems, construction errors, environmental incidents, and/or human interference. The consequence can be either serviceability failure or ultimate strength failure. A serviceability failure occurs when a pipe does not collapse, but rather fails to meet the required specifications. For example, pipeline leakage can be considered as serviceability failure, because the pipe does not operate well but it is still in place. In ultimate strength limit state cases, a pipe completely loses its structural integrity and ends up in ultimate collapse. While cracking and leakage can be in the category of pipeline serviceability failure, flexural failure, and shear failure are involved in the category of ultimate strength failure.

Among the causes of pipe failure, corrosion is often the most significant. A study over the period of 1999−2001 of the percentage distribution of all failures in the Unified gas supply system showed more than 44% of failures are related to pipeline corrosion (Litvin and Alikin, 2003), see Fig. 2.1.

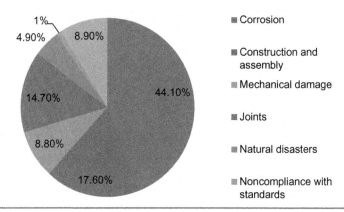

Figure 2.1   Causes of gas pipeline failure.

To detect and prevent these failures, operators, and regulators conduct regular inspections. Inspections generally target typical pipeline vulnerabilities in three categories: corrosion, deformations, and cracking. Corrosion may be the most consistent integrity challenge facing operators of cast iron and steel pipelines. While other concerns can be considered incidental and only affect certain sections of pipeline, corrosion constantly affects every inch of pipeline.

2.2.1 Causes of Pipeline Failure

Corrosion: Corrosion is a natural phenomenon that occurs due to exposure of the pipe to the surrounding environment. Corrosion mechanisms for each type of pipe were mentioned in Section 1.4 of this book.

Left unchecked it can eventually degrade the structural integrity of a pipeline. Corrosion generally results in minor leaks from small holes in the pipeline. Fig. 2.2 shows corrosion progress of a painted steel pipe in industrial facilities in Taiwan.

Improper maintenance of pipelines can lead to stress-induced fracture or cracking. Stress is a physical quantity measured in Pascal that describes the force per unit area acting on a material. Stress generally leads to cracks in pipelines in three ways: cyclic fatigue, stress corrosion, or manufacturing error. Cyclic fatigue is the structural damage that occurs when the pipeline is subjected to fluctuating internal pressures. Stress-corrosion cracking occurs where the pipe is under tension and exposed to corrosive elements. Cracks that are built into the pipe tend to be too small to cause pipe failure but are usually detected nevertheless. Cracks generally cause leaking but severe cracks can lead to a burst pipe.

As defined by the Oxford Dictionary to corrode is, "to destroy or damage (metal, stone, or other materials) slowly by chemical action" as explored

Figure 2.2 Corrosion of a steel pipe in a factory in Taiwan.

previously in Section 1.4, for each material type reacts differently to different chemicals, whether it be the substance contained within the pipe or the environment surrounding the pipeline itself.

Clay is a very brittle substance, but has proven to withstand the natural and aggressive elements of soils in which the pipelines are buried. But substances flowing through these pipes need to be chosen very wisely to prevent corrosion from occurring.

Concrete pipe corrosion occurs predominantly due to the reinforcement on the inside. Once the steel reinforcement is exposed to oxygen, rust starts to form and causes further deterioration. Sulfide corrosion in concrete sewers can also be prominent.

Metal pipelines are most vulnerable to corrosion due to rust as explained previously. Also metals react with some gasses and at high temperatures, which accelerates corrosion.

Corrosion in plastic pipes occurs due to heat exposure, this changes the structure of the plastic thus causing it to deform, thinner walls of the pipeline can be the consequence. This increases the likelihood of cracks, leakages, and ruptures.

Third party interference: One of the common causes of deformations in pipelines is the human interference. When heavy equipment or rocks strike the pipe, they can make dents or gouges that change the internal geometry of the line. Changes in the internal geometry alter the distribution of internal pressure by focusing it on certain sections of pipeline. Over time these deformations may result in pipeline failure. Deformations are often accompanied by a loss of coating, which increases the risk of corrosion.

Because human interference is the main cause of deformations, preventative measures predominantly target public awareness. Pipeline operators participate in federal, state, and provincial programs in both the United States and Canada provide toll-free hotlines for property owners planning any excavation projects. Property owners are required by law to check with these programs before they begin a project. To increase the general visibility of submerged and buried pipelines, land-based pipelines are often clearly marked and no-anchor zones are established along aquatic pipelines. Federal law requires that markers be located "at each public road crossing, at each railroad crossing, and in sufficient number along the remainder of each buried line so that its location is accurately known."

With regards to pipelines, third party interference is a very damaging aspect. As its name implies this means that there is a contributing third party to the deterioration or failure of the pipeline. This can be in the forms of unauthorized or mechanical digging, excavation, etc. The more exposure pipelines have to these third parties the higher the likelihood of failure thus reducing the reliability of the pipeline significantly.

Landslide: Land movement can be one of the easiest ways to decrease pipeline structural reliability; especially movements as significant as a landslide or earthquake. In the design stage of pipelines, natural disasters can be taken into account, especially if they are expected within the risky area.

2.2.2 PIPELINE FAILURE MODES

Failure modes can be presented mathematically by using the "limit state" concept. A limit state is a condition of a structure beyond which it does not fulfill the relevant design criteria. The general formulation for limit state function of a structure subject to a time varying process (e.g., corrosion) is presented as follows:

$$G(R, S, t) = R(t) - S(t) \qquad (2.1)$$

In the above equation $G(R, S, t)$ is a limit state function (or failure mode), $S(t)$ is the stress affecting the limit state requirement at time t, and $R(t)$ is the resistance of the structure at time t. The failure occurs when the stress becomes greater than the resistance of the structure. More details about the limit state concept in reliability analysis can be seen in Chapter 3, Methods for Structural Reliability Analysis.

There are many known failure modes for pipelines. Identifying the dominating failure modes depends on the definition of the physical model for the system, which involves consideration of loads or any other contributing parameters. Traditionally the most important parameters involved in the analysis of pipes are loads, pipe mechanical properties, pipe geometry, and soil characteristics in the case of buried pipes (Moser and Folkman, 2008). External loadings and corrosion, which acts through reduction of the pipe wall thickness, affect the failure condition of the pipeline.

Several limit states (failure modes) can be considered in the failure analysis of pipes. Normally the limit states are those which are controlled in designing stage of pipeline (Moser, 2010).

Longitudinal deflection: Nonuniform soil compaction along with overexcavation can be the reason behind nonuniform bedding, which leads to longitudinal deflection of the pipe.

Pipeline deflection is considered as a serviceability failure, with the limit state function presented as follows:

$$G(Y, Y_{max}, t) = Y(t) - Y_{max} \qquad (2.2)$$

If the deflection (i.e., $Y(t)$) exceeds the allowable longitudinal deflection threshold (i.e., Y_{max}), the deflection failure will happen. Longitudinal deflection can be calculated by Eq. (2.3), (Gorenc et al., 2005).

$$Y = \frac{P_s L^3}{250 \text{EI}} \qquad (2.3)$$

The critical value for longitudinal deflection suggested by Gorenc et al., 2005 is as follows:

$$Y_{max} = \frac{5L^3}{384 \text{EI}} \qquad (2.4)$$

Ring deflection: It is also necessary to study the ring deflection of pipe and make sure it does not reach 5% of inside diameter of pipe to prevent ring deflection failure (Moser and Folkman, 2008). Ring deflection also can be considered as a serviceability failure with the limit state function presented as Eq. (2.5).

$$G(\Delta X, D_i, t) = \Delta X(t) - 0.05 D_i \qquad (2.5)$$

If the ring deflection (i.e., $\Delta X(t)$) exceeds the allowable threshold ($0.05D_i$), the ring deflection failure will happen. Ring deflection can be calculated by Eq. (2.6) (BS 9295, 2010).

$$\Delta X = \frac{K(D_L W_c + P_s)D_m}{\frac{8EI}{D_m^3} + 0.061E'} \tag{2.6}$$

The allowable threshold for ring deflection suggested by BS 9295, 2010 is 5% of inside diameter of the pipe.

Leakage: Leakages reduce the reliability of a pipeline significantly, because a leak can lead to a rupture if not identified and treated appropriately. Leakage happens when the depth of corrosion pit exceeds the pipe wall thickness. Leakage can be considered as a serviceability limit state and the formulation for the limit state function can be presented as Eq. (2.7).

$$G(\Delta, W_t, t) = \Delta(t) - W_t \tag{2.7}$$

In which $\Delta(t)$ is the depth of corrosion and W_t is the critical value for leakage which is the pipe wall thickness.

Buckling: Buckling can be a major form of reliability reduction mostly for plastic pipelines, due to having more flexible structure in comparison to steel, concrete, and clay. Buckling affects the reliability of a pipeline due to the stretching of the pipeline which leads to a thinning in the pipe wall, and consequently thinner walls reduce the structural reliability.

Buckling as a serviceability limit state is also considered as one of the critical failure modes which should be kept less than its critical thresholds to guarantee the safety of pipeline.

Fig. 2.3 shows an above ground steel pipeline. Loading and environmental conditions caused the pipeline to buckle and reduce its reliability, although the pipeline hasn't failed, there is a vital need to repair.

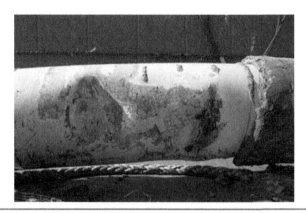

Figure 2.3 Example of buckling failure in a steel pipeline.

Buckling limit state can be presented in the form of Eq. (2.8).

$$G(P, P_{cr}, t) = P(t) - P_{cr} \qquad (2.8)$$

in which buckling strength (i.e., $P(t)$) can be calculated by Eq. (2.9) (Moser and Folkman, 2008).

$$P = \frac{1}{S_f} \sqrt{\left(32 R_w B' E_s \frac{EI}{D_m^3} \right)} \qquad (2.9)$$

and the critical value for buckling suggested by Moser and Folkman (2008) is as Eq. (2.10).

$$P_{cr} = R_w \frac{W_c}{D_m} + \frac{P_s}{D_m} \qquad (2.10)$$

Flexure: Excessive bending can result in flexural failure which is an ultimate failure. If the flexural stress on the pipe exceeds the flexural strength, pipe fails structurally. Flexure limit state function can be presented in the form of Eq. (2.11).

$$G(M_n, F_y, t) = M_n(t) - f_y \qquad (2.11)$$

For steel pipes the equation to calculate flexural capacity suggested by Gabriel (2011) is as follows:

$$M_n = \frac{2 D_f E \Delta Y y_0 S_{f1}}{D_m^2} \qquad (2.12)$$

The critical threshold for flexure failure is f_y which is yielding strength of pipe material.

Shear: If a pipe loses its shear strength it is completely failed. Therefore shear failure mode is considered as the ultimate strength limit state function, which can be presented as Eq. (2.13).

$$G(V_b, V_s, t) = V_b(t) - V_s(t) \qquad (2.13)$$

The formulation for shear strength for a reinforced concrete pipe suggested by ASCE 15−98, 2000 is as follows:

$$V_b = 0.083 b \varnothing_v d F_{vp} \sqrt{f_c'} \left(1.1 + 63 \times \frac{A_s}{bd} \right) \left[\frac{F_d F_N}{F_c} \right] \qquad (2.14a)$$

$$F_N = 1 + \frac{N_u}{3.5 bh} \qquad (2.14b)$$

If the shear stress (i.e., $V_s(t)$) exceeds the shear strength of a pipe (i.e., $V_b(t)$), then failure happens.

Wall thrust: Wall thrust failure is an ultimate structural failure with a limit state function presented in Eq. (2.15).

$$G(T_a, T_{cr}, t) = T_a(t) - T_{cr} \tag{2.15}$$

The pipe capacity can be calculated using the following equation (Gabriel, 2011):

$$T_a = F_y(W_t - \Delta)\varphi \tag{2.16}$$

The threshold can be calculated from:

$$T_{cr} = 1.3(1.67P_sC_L + P_W)\frac{D_o}{2} \tag{2.17}$$

Fracture toughness: For localized stress concentration caused by defects, e.g., corrosion pits, a term stress intensity factor, K_I, is used in fracture mechanics to more accurately predict the stress state ("stress intensity") near the tip of a crack (caused by applied or residual stresses). It is a parameter that amplifies the effect of stress field at the tips of crack leading to fracture. In essence, K_I serves as a scale factor to define the magnitude of the crack-tip stress field and is related to the geometrical parameters and stress types of the element (Hertzberg, 1996).

In general, there are three deformation modes of fracture (Hertzberg, 1996): (1) opening mode (Mode I); (2) in-plane shear mode (Mode II); and (3) out-of-plane shear or tear mode (Mode III). Since Mode I is found to be the dominant cracking condition in pipes under normal service conditions (Laham, 1999), most of the time, only Mode I is considered in the failure analysis of the pipe (i.e., a crack plane is perpendicular to the direction of the stress incurred).

If K_{IC} is the critical stress intensity factor, known as fracture toughness, beyond which the pipe cannot sustain propagation of the crack pit, the two limit state functions for fracture toughness can be established as follows (Mahmoodian, 2013):

For axial fracture: $G(K_{IC}, K_{I-h}, t) = K_{IC} - K_{I-h}(t)$ (2.18)

For hoop fracture: $G(K_{IC}, K_{I-a}, t) = K_{IC} - K_{I-a}(t)$ (2.19)

In Eqs. (2.18) and (2.19) it is assumed that fracture toughness of the pipe material is the same for both hoop and axial directions.

Laham (1999) presented a formula for stress intensity factor for a crack pit in a pipe under hoop stress as follows:

$$K_{I-h} = \sqrt{\pi a} \sum_{i=0}^{3} \sigma_i f_i \left(\frac{a}{d}, \frac{2c}{a}, \frac{R}{d} \right) \tag{2.20}$$

and the stress intensity factor for a crack pit in a pipe under axial stress:

$$K_{l-a} = \sqrt{\pi a} \left(\sum_{i=0}^{3} \sigma_i f_i \left(\frac{a}{d}, \frac{2c}{a}, \frac{R}{d} \right) + \sigma_{bg} f_{bg} \left(\frac{a}{d}, \frac{2c}{a}, \frac{R}{d} \right) \right) \qquad (2.21)$$

For internal and/or external crack pits, the difference in formulations of stress intensity factor (Eqs. 2.20 and 2.21) lies in geometry functions (i.e., f_i and f_{bg}), which have been presented in different tables by Laham (1999). Due to the propagation of corrosion, crack depth (i.e., a) and crack length (i.e., $2c$) change with time so the stress intensity factors are time variant.

2.3 Inspection Methods

The majority of current pipeline inspection methods involve the use of manual labor and human evaluation. The main concern is that it requires large amounts of time to complete the assessment of defects. In addition, human error is expected in the evaluation which in some cases result in inaccurate assessment. This has encouraged engineers and researchers to innovate and develop potential solutions for inspection of pipeline networks. Pipes run for hundreds of kilometers and in many cases they are underground. Therefore the innovative methods for assessment can be automatic inspection systems. The merit of those systems is that they greatly reduce the time taken to properly inspect pipes over long distances and also if applied correctly, can eliminate the human errors.

In this section, conventional methods and some of the most recent advanced methods for pipeline inspection are presented.

2.3.1 CCTV Method

Currently, the most common method used to investigate the internal condition of pipes and perform defect assessment is known as closed circuit television (CCTV). There are a number of leading worldwide companies who produce a diversity of CCTV modules to accommodate a range of pipe sizes and scales of work.

CCTV is relatively cheap and identifies easily physically damaged pipeline walls or pipelines under threat of corrosion, crack, or deformations. The technology allows engineers to decide whether to act further on the pipeline structure, conduct more testing to assess the wellness of the pipeline structure, relocate the pipeline system, or replace it entirely.

In a CCTV device system, a remote controlled car is placed inside the pipeline and is used as a visual inspection tool to view the inside of the pipe

Figure 2.4 Closed circuit television for pipe inspection.

without the need for extraction of the pipeline (Fig. 2.4). This means of inspection is cost-effective and reliable in identification of defects at the assessed areas. The camera on the front can be rotated during the course of the navigation within the pipe to identify more than just what is directly in front of the car. The machine is collaborated within the pipe and exact coordinates and distances are present in the footage to enable identification of damage within a pipeline.

Damaged or defected pipes, generally can be treated in three ways: cleaning, repairing, or replacing. This is all determined once the report has been provided to the pipeline engineers/maintenance team from the contracted CCTV inspection company. Once the reliability of the pipeline is determined from the testing, the maintenance strategies are enforced to ensure the structure can rerun at its highest efficiency.

Visual data from the CCTV devices is transmitted via cable in real time to the operator and is recorded for further in-depth analysis in photo or video formats. The CCTV is capable of detecting notable damage or defects within the pipe, which is the subject of stream quality and fully relies on operator competency to correctly record and classify the defect (Ahrary et al., 2007). Considering a number of influencing factors some research (Korving, 2004; Korving and Clemens, 2005a,b; Dirksen et al., 2013) concluded that CCTV data may lead to defect misidentification by an error rate as great as 20%−30%.

On its own, visual inspection produced by a CCTV module provides factual evidence of certain pipe defects, i.e., the presence of a defect, its severity, and longitudinal and circumferential location. However, time-consuming supportive hardware and/or software is required to carry out an accurate quantitative assessment of the scale or depth of a given defect. In this instance, a number of supportive methods were developed to facilitate the numeric interpretation of the CCTV data (Duran et al., 2007; Sarshar et al., 2009; Hao et al., 2012; Rogers et al., 2012), however the direct use of measuring defects geometry has not been mentioned.

The process of CCTV involves the human operators continually playing and stopping the camera footage to determine regions of interest (ROI). ROI are defined as those areas where potential defects may occur. After determining these areas, the operators must then make a decision as to whether further examination is required for that particular image.

Even collecting large amounts of data and visual images, the pictures are still viewed by human operators and therefore the error prone factor of the inspection is still a concern for CCTV inspection.

Fig. 2.5 presents the types of defects which professional CCTV operators who are responsible for identifying defects are looking for. These defects are in accordance with Pipeline Assessment and Certification Program (PACP).

Until a proven automated method of assessment is developed, we can only improve the quality of evaluations by improving the human operators through experience, knowledge, and training in specific working conditions to acquire the best possible outcome.

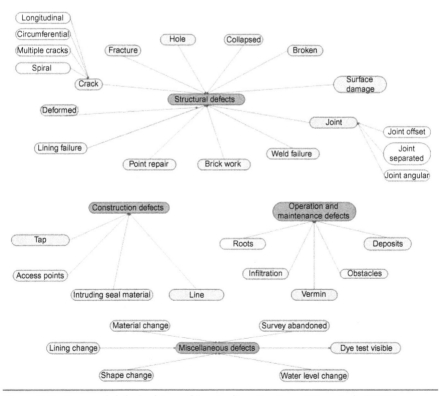

Figure 2.5 Types of defects detected in pipeline structures (Guo et al., 2009).

2.3.2 LASER SCANNING

Another inspection method which is similar to the CCTV, is the laser scanning (Fig. 2.6). CCTV produces high quality imaging however, the laser scanning provides even greater detailed pictures, which is greatly beneficial and improves the effectiveness of successfully identifying defects in pipeline networks. Laser scanning is able to demonstrate segmentation and classification of some common defects within pipes such as cracks, joints, and holes. This technology is not widely used among the industry due to its high inspection costs, time taken to complete an evaluation of pipes, and other implementation limitations.

Laser scanning together with supportive software quickly spread out across pipeline inspection techniques in 2000. Duran et al. (2007) describe in their study how the CCTV camera device can be mounted with a laser-based profiler and an optical diffuser and is supported by a defect classification algorithm. The algorithm was based on the image intensity values quantification of the projected rings viewed by the camera where intelligent image processing and Artificial neural network (ANN) techniques were adopted. This approach enables the building up of a signature database for a number of defect types for future automatic identification. Conventional laser scanning techniques have an incorrigible error in measurements which is equal to the diameter of the laser spot, usually around 4 mm. An advantage of the method proposed by Duran et al. (2007) over other existing methods is that laser size has no implication on result accuracy and even a small imperfection can be measured; as such, the results were reported to have an accuracy of 1 mm. Additionally, the proposed method has no significant requirement to have the laser ring pointing directly in the pipe center and does not require an extra smooth motion.

Figure 2.6 Laser scanner for pipe inspection.

Ahrary et al. (2007) report on a novel KANTARO robotic system that is designed to detect faulty areas in sewer pipes, based on image acquisition and 2D laser readings. The robot uses a combined graphic user interface and intelligent fault detection module to detect various sewer defect types automatically, which can be reported to the user in real time. The machine is designed to inspect 200–300 mm pipes with 90° bends and maximum steps of 50 mm. The authors reported an accuracy of ± 1 mm; however, no information is provided as to whether the measurement accuracy is applicable to the lateral/circumferential distance of size/depth of defects.

The rotating optical geometry sensor (ROGS) was developed by IOSB, Germany (Ritter and Frey, 2010) and was specifically designed to survey buried pipes. The sensor consists of four straight laser beams arranged at 90° angular displacements and an optical triangulation technique is adopted to perform the wall scan. This technique allows for reconstruction of a pipe wall surface in 3D or plain regimes. The laboratory and field experiments in clean water and sewer pipes revealed that an accuracy of pipe wall discontinuity of 1 mm can be achieved.

Stanić et al. (2013) investigated the means of egg-shaped concrete sewer pipe interior and exterior scanning using a laser-based profiler. The obtained images from inside and outside of the pipe were used to reconstruct a 3D model of the pipe wall surface, with further ability to measure the wall thickness and thinning rate. The code to perform the operation was developed in MATLAB which also enabled measurement of the pipe wall roughness coefficient based on Darcy-Weisbach and Chezy equations. The results revealed that with the laser profiler there is an average measuring error of 0.33%, which was approximately 1 mm of the pipe wall thickness. However, distance measurement could have an error of 3 mm. When combining the results, a common error of 4 mm may be present. The technique offers good accuracy for overall pipe wall condition review, although it would require improvement if it was to be used in the field to detect corrosion rate.

2.3.3 ULTRASONIC INTELLIGENT PIGGING

Intelligent pigging allows for a pipeline to be inspected for faults such as cracks, corrosion, rust, deformations, etc. The method uses a range of nondestructive techniques such as ultrasonic and magnetic leakage testing and also allows the pipeline to be cleaned.

The magnetic method utilizes applications of a magnetic field within the interior of the pipe wall using permanent magnets. An irregularity in the magnetic field due to varying wall thickness will be detected by the sensors and therefore can be used to determine quantitative data and presence of defects in pipes. More details of the magnetic method are explained in Section 2.3.4.

The ultrasonic method is able to obtain quantitative data relating to the depth of defects with precision to the millimeter. This process is based on propagation delay time of an acoustic wave emitted by ultrasonic transducers that travel in a liquid medium and reflected by internal pipe walls to detect and evaluate corrosion within pipes. The types of defects that can be identified depends on the characteristics of the proposed ultrasonic transducer that is used, the precision of time measurement of propagation of acoustic wave, and synchronization of transducer's trigger with measurement of the pig's movement.

The pigs normally have a mechanical part which consists of a cylindrical capsule supported by rubber discs. The capsule rotates freely around its longitudinal axis to keep ultrasonic transducers directed to the lower half of the pipe. The function of the rubber discs is to keep the capsule centered, block fluid, and propel the pig forward in the inspection process. An odometer wheel is also held outside the capsule to keep track of the total distance covered by the pig whilst inside the pipe by generating pulses every 10 cm traveled. The pig uses a number of ultrasonic transducers in a mechanical support in the capsule's plug. The transducers' function is to measure echo time propagation. Transducers are kept away from internal pipeline wall by implementation of mechanical support and rubber discs. A microcomputer is connected to the pig for the calibration process before inspection begins. The beginning of inspection is detected by generation of pulses from the odometer wheel. Once pulses are detected again the inspection will resume. This process is designed to optimize energy consumption while the pig travels extremely long distances of pipeline infrastructure in order to save battery power for more important uses. The pig can acquire 250 measurements per second per transducer, however, the electronics are developed to cope with a rate of 500 measurements per second per transducer. If there are 16 transducers traveling with the pig, then 4000 measurements per second are able to be collected as data acquisition. The echoes detected by the pig classify defects at certain depths and the classification is based on the percentage of corrosion within internal walls. While data is being collected by the pig, the software uses a data compression technique to save storage space as space is limited on the disc. The basic principle of this mechanism is that there is no purpose in storing data and information for pipelines when no defects are present. Improved axial and lateral resolutions can be achieved by increasing the frequency among transducers since it produces shorter wavelengths.

The flow of product can continue whilst routine checks and cleans are occurring, allowing for minimal disturbance to be caused. Fig. 2.7 depicts the intelligent pigging device.

Figure 2.7   Example of an intelligent pigging device.

A three-phase automated pipe condition assessment is normally considered while using an intelligent pigging device, including defect detection, defect interrogation, and defect classification.

The first phase of the process (defect detection) involves the use of a robot which moves at a relatively normal speed to detect defects along the pipeline. If a potential defect is identified, the robot stops and records higher quality visual images and data to determine either it is a real defect or just a false alarm. After this step, the robot will continue to move forward and look for more defects.

The second phase (defect interrogation) involves the robot stopping completely once a defect is determined, and performing an examination of the area where it was detected. Here, the robot will discriminate each defect specifically among several types such as cracks, fractures, roots, corrosion, or lining failure.

The third phase (defect classification) involves classifying the defects either during the inspection or online after the entire inspection process has been completed using several classification techniques which the system has been programmed with. For example, if a crack was identified in the second phase, it could be identified as a horizontal or spiral crack in the third and final phase of the inspection. The final classification also determines the severity of the defect and whether it needs immediate attention, further monitoring or it is safe to ignore it at the current time.

2.3.4 MAGNETIC FLUX LEAKAGE TECHNIQUE

Magnetic flux leakage (MFL) detection technique is one of the common methods of pipeline inspection in oil and natural gas industry. It is a nondestructive testing technique which uses magnetic sensitive sensors to detect the magnetic leakage field of defects on both the internal and external surfaces of pipelines. It provides very high and accurate resolution pictures of the interior walls of pipelines where defects and other anomalies can be identified and assessed.

This in-line inspection (ILI) technique magnetizes the pipe wall as it travels down the structure and uses coil sensors to measure magnetic flux leakage intensity along the pipe walls. MFL is based on an array of sensors placed along the interior of the pipeline walls which detect changes or irregularities in the magnetic field. These irregularities are where potential defects may exist and can be evaluated to determine what type of defect is present. The severity of the defects is determined by estimated defect depth and consequently the safety of the pipe by calculating maximum allowable operating pressure (MAOP) of oil or gas through the pipes. Some common defects that this technique may identify include corrosion, deformations, fatigue, hairline cracks, dents, buckles, delaminations, and faulty welds.

Fig. 2.8 schematically illustrates how the sensors are placed on the pipe wall and how the defect/crack can be detected. The tool includes the use of magnets and steel couplers which touch the pipe wall to create a magnetic circuit. If the wall thickness increases or decreases at a particular section, magnetic flux change will be detected which allows the defect to be identified and later be evaluated through the data acquired.

Since MFL technique can record changes in more than one or two directions, it can detect even the small defects by most subtle changes of magnetic. After the inspection is completed, using the MFL technique, data and image are transferred to a computer monitor where the information is rendered and stored as 2D or 3D images of the exterior or interior section of pipe walls.

Using the MFL technique, various submethods of pipeline inspection were used for classification and regression of the infrastructure. The methods used included Kernelization, regularized least-squares regression, support vector machines for regression, kernel PCA, and partial least squares regression. The goal of these procedures is to extract the most relevant features from a set of candidate features.

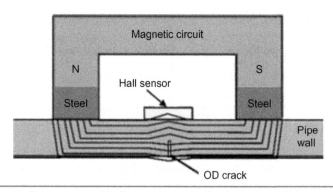

Figure 2.8 Performance mechanism of magnetic flux leakage technique.

Machine learning algorithms are used in the analysis of the MFL images. These algorithms look at two primary issues:

- Binary classification of image segments into injurious or metal loss defect, and noninjurious safe or nonharmful objects.
- Estimating the severity of defects.

2.3.5 RADIOGRAPHIC TESTING

Radiographic testing (RT) is carried out by recording degree of absorption of penetrating radiation throughout the pipe wall. This produces a latent image of the object being examined on a film which is then chemically processed to transform latent image into a permanent shadow image of internal and external conditions of the object. A greater amount of radiation passes through a defected area compared to a region without defective issues. These radiographic images can be evaluated by either human operators or the automated computer vision system currently being tested for accuracy.

Radiography film has been the foundation of NDT for the last 50 years. As discussed previously, manual inspection methods and similar procedures are extremely time-consuming and often cannot be 100% accurate or reliable.

A large number of radiographic images are required over the long distances of pipeline networks and this makes the identification process of welding defects very time-consuming. Additionally, it is even more challenging as human operators have to assess and evaluate the large number of images and often people will have different opinions regarding defects. Developments in digital image processing and computer vision have allowed the extensive use of automated visual inspection to be tested and applied more frequently since it is much more consistent and effective in terms of operating compared to human operations. It also allows the inspection of pipelines in unsafe conditions to be analyzed in more detail. The automated computer vision system utilizes the use of radiographic films of welded pipelines which are produced by a radiographic testing.

2.3.6 ACOUSTIC DETECTION

Lohr and Rose (2003) spotted the potential of using ultrasonic guided circumferential and longitudinal waves propagating in a pipe wall to detect the pipe and load properties. For guided waves in a circumferential experiment, it was shown that the amplitude of the received signal decreases as the properties of the load inside the pipe (air, oil, water, vegetable shortening) or wall thickness changes. Alterations of up to 1 mm could be recorded. In longitudinal experiments, it was found that changes in pipes, from clean condition to having tar levels of just 0.25 mm, are detectable and also results in a decrease of received signal amplitude. However, this method is yet to be explored in terms of its potential for different type of pipe applications.

A challenging field based test was undertaken by Lewis and Fisk (2005), where they used the sonic/ultrasonic method to measure a 152 cm concrete pipe wall which was thinning due to a number of factors. For this purpose test pits were excavated along the pipe to allow four sonic sensors based on a curve to be placed on the pipe outside the coating. Unfortunately, no information on the sensor type and processing technique was provided. However, the results showed that recorded resonate frequencies of 16 to 18 kHz corresponded to pipe wall thicknesses of 11.2−11.7 cm and frequencies as high as 21−29 kHz indicated that the pipe wall thickness was reduced. Unfortunately no assessment of method accuracy, comparison, and further application was suggested.

The above research was carried on further and Fisk and Marshall (2010) published extended field study results, where 81 prestressed concrete cylinder pipes (PCCP) all of 122 cm diameter were assessed either from inside, having the pipes fully drained, or from outside by digging up test pits. The sonic/ultrasonic sensor identified 26 pipes with wall thickness loss, as the resonate frequency shifted. For 122 cm pipes with a thickness of 10 cm, the normal resonate frequency was 20−24 kHz; however, for thinner walls measurements of 29−30 kHz were recorded. To validate the findings, the 26 pipes were excavated in those places where the sonic/ultrasonic measurements had detected wall thinning; all findings were confirmed. However, no quantitative measures were provided.

Horoshenkov et al. (2004, 2011) developed a quick and low-cost acoustic method of pipe inspection using a sinusoidal chirp sound transmitted in the range 100 Hz−20 kHz and applying methods of signal reflectometry, sound attenuation and Fourier Transform standard deconvolution. The acoustic sensor consists of a speaker and a microphone array which are attached to a pole that is stationed in a manhole below the pipe soffit. The cable from the sensor is attached to an electronic box leading to a portable computer from where the sensor is activated; the survey results are instantly displayed and stored. Undertaken sets of laboratory and field trials on clay, concrete, PVC, and Perspex pipes of various diameters, and including a number of defects, conclude that the acoustic sensor is able to detect large defects such as blockages, roots, missing pipe segments, cracks, fat build-up, and sediment, as well as small pipe wall permeability contrasts such as microcracks, encrustation, and pipe wall roughness (as corrosion). It was reported that the software defect recognition "signature" database can be built up and trained; however no physical measure of the condition scale is reported.

2.3.7 GROUND PENETRATION RADAR

Ground Penetrating Radar (GPR) is a well-established method of monitoring civil structures such as bridges, tunnels, and buildings. GPR is also used as part of a nondestructive detection method to identify underlying features or buried services. It is well known to be a reliable method of identifying the location and depth of underground pipes of different materials and sizes.

Koo and Ariaratnam (2006) carried out field studies, using GPR to identify internal sewer damage in concrete and clay test pipes of 5−50 cm in diameter, followed by concrete and PVC field pipes of 75−90 cm in diameter. The method reported to have a number of limitations due to its bulky size and the length of antenna; however studies revealed that even small defects, such as material loss due to deterioration and corrosion, can be identified (no quantitative figures were provided). In general, compared to other techniques, GPR is capable of detecting geometric discontinuities and changes in material properties that the human eye would be unlikely to notice.

Digital scanning and evaluation technology (DSET) is a proven method to provide a high definition quality of visual data. Koo and Ariaratnam (2006) combined this technique with GRP to support the measurements. The results of the merged techniques proved a great correlation between radar and visual data which helps to classify defects in a consistent and error-free manner.

2.3.8 METHODS OVERVIEW

A number of inspection techniques which are specially designed to monitor the pipeline condition or those that have a great potential to be further developed in this context were reviewed in the previous sections. A summary of these techniques including their advantages and disadvantages is presented in Table 2.1.

2.4 Pipeline Maintenance

A variety of preventative measures can be taken to reduce corrosion of pipelines. Methods include implementing cathodic protection, regular cleanings, and pipe coatings. The simplest and most effective way to prevent a pipeline from corroding is to keep it out of contact with the environment.

To ensure pipelines are operating at their highest functionality routine maintenance must be undergone. Depending on the age and material of the pipes

TABLE 2.1 Comparison of Pipe Inspection Techniques

Method	Description	Advantages	Disadvantages
CCTV	Uses hard wire push systems and crawlers/tractors to drive the camera through sewer pipe. Inspection is based on operator decision or complex software	• Visual provision of pipe condition • Effective in dry test or with low flow • Compatible with autonomous fault detection systems • Intelligent fault detection reduces inspection time • Fault detection software reduces time spent on pipe condition review and demonstrates consistency between image quality and in-pipe visibility • All types of known and new defects can be detected	• Survey and review process is time-consuming • Picture quality affects operator's decision (is subject to illumination, humidity, etc.) • Quantifying the defect size is a big challange • Tractor mobility is constrained by sediment, debris, hard compacted solids, and high flow levels • Requires development to operate in live pipes • Defect classification relies on operator competency • Defect misidentification by operator is high • High cost of supporting software
Laser scanning	An external laser beam projector is placed on top of CCTV equipment. Special software is required to interpret the data	• Easily incorporated into existing CCTV systems • Automated defect classification algorithm • Estimation of pipe wall roughness/loss • Cracks, pipe joints, and sedimentation can be detected • Image automate processing using ANN technique • Single inspection run • 360° scanning angle	• Depends on sharp images of projected laser ring • Depends on CCTV crawler smooth mobility • ANN technique requires defect signature database • High cost of camera and laser • High cost of comprehensive software • Time-consuming survey
Ultrasonic intelligent pigging	Based on propagation delay time of acoustic wave emitted by ultrasonic transducers	• Ability to inspect long distances in short time • Very high and accurate resolution images	• High cost of the instrument • Data storage problem

(Continued)

TABLE 2.1 (Continued)

Method	Description	Advantages	Disadvantages
Magnetic flux leakage	Uses magnetic sensitive sensors to detect the magnetic leakage field of defects	• Ability to inspect long distances in short time • Very high and accurate resolution images	• High cost of the instrument • Data storage problem
Radiographic testing	Recording degree of absorption of penetrating radiation throughout the pipe wall	• Can detect defects specially in welds	• High care during the radiography (as it is harmful for humans) • Huge number of images need to be analyzed • Time-consuming process • Human errors
Acoustic detection	Uses an array of microphone and a speaker. Usually placed in a single position to survey the whole length of the pipe. Requires software to interpret data	• Rapid measurements • Guided waves differentiate between materials • Acoustic velocity or attenuation impact can be used to estimate wall corrosion • Simultaneous use of methods allows for more precise prediction of defect severity • Pipe wall thickness can be ascertained • Compact and portable equipment to fit all pipe sizes • Low-cost technology • Automated defect classification • Easy to understand instant pipe survey report • Detects all known pipe defect types	• Some field prototypes are unsuitable for small pipes • Some field prototypes require ground excavation up to pipe crown • Some software requires the operator to be trained and hold a defect signature database • Software varies in price and may be expensive
Ground penetration radar	Uses point source emitting pulsate electromagnetic radiation and a receiver which translates the data into spectrogram	• Capable of indicating moisture in pipe walls and potential leakage spots • Allows the estimation of pipe wall thickness • Able to detect pipe deterioration	• Bulky equipment • High equipment cost • Software costs vary and may be high • Requires special training to interpret data

determines the frequency which the routine maintenance methodologies need to occur; naturally the older the pipeline and the more brittle and easily corroded the pipeline, the more dependent on maintenance.

2.4.1 COATINGS

Any pipeline that is buried or submerged is required by standard specifications to have an external coating for external corrosion control. Many pipeline operators use an epoxy coating to seal off the surface of the pipe. While epoxy coatings are the most popular, the diverse terrain through which pipelines operate sometimes require specialized coatings. For example, a pipeline operator might use a cement coating for a pipeline crossing a river to help weigh it down and keep it in place.

There are different means of coatings specific to the original material in which a pipe is made from as well as the purpose it serves. Coatings can be for the inner or outer surface of the pipeline, and some are more specific to the pipeline joints. There are variety of coatings commercially available to assist in the protection of existing pipeline infrastructure from corrosion.

The coatings can come in spray forms, putties, tapes, and paints. Coating of pipelines creates an additional barrier round the pipe to protect it from the substance running through the pipeline as well as the corrosive substances found in the soil which the pipeline is surrounded by, thus adding to its functional integrity.

Epoxy coating is one of the most commonly used maintenance techniques used for pipelines. Epoxy coating shields the pipelines base material from the substance which the pipeline is transporting as well as previous damage to the pipe. Within a steel pipe, rust is almost inevitable especially when used in water transportation. Rust isn't desirable especially if the water being transferred is for drinking; the rust can be harmful to humans. Once the pipe is assessed and cleaned, the epoxy coating can be applied to prevent further rusting to occur, preventing complete pipeline failure, and allowing the water to be much more sanitary. In the terms of a concrete pipe, epoxy coating is used to shield the pipe from sulfide corrosion as well as fill in cracks which may cause pipeline failure.

Plastic coating on pipes (more specifically metal pipes) is also common. This arrangement of coating prevents external damage to the pipe from the surrounding environment. Ground water, salinity, and other minerals within the grounds' surface can be damaging to pipelines. The plastic coating creates a thin barrier between the pipe material and corrosive substances thus preventing the pipeline from deteriorating externally. This prolongs the pipeline's life span as well as its reliability.

2.4.2 CATHODIC PROTECTION

Cathodic protection is another common method for protecting steel pipelines from corrosion. In older and simpler galvanic systems this involved coating the pipeline in an anode, such as zinc, to prevent the steel from reacting to corrosive environment. The anode corrodes in place of the steel. Because the zinc eventually corrodes and leaves the steel bare, galvanic systems have a limited lifetime. Newer pipelines are required to make use of an impressed current protection system (Fig. 2.9). In this system, the steel pipeline is connected to an anode made of a metal that is more reactive than steel (e.g., magnesium, aluminum, zinc). Because the anode is more reactive, it needs to lose all its ions before the steel begins to corrode. The anode is then connected to a power source called a rectifier. As the anode loses electrons to corrosion in place of the steel, the electrons are replenished by the power source. This system is limited only by the rectifier and therefore has a much longer lifespan.

Steel pipeline maintenance can be undergone by cathodic protection. Water and fuel pipelines which are made of steel are the most common to undergo this type of protection.

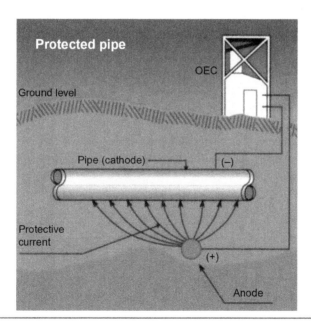

Figure 2.9 Cathodic protection of pipes.

2.4.3 CLEANING

Regular cleaning is an industry standard for oil pipeline operators. These cleanings ensure that the pipeline is operating at peak efficiency and that corrosive buildup is removed. Pipelines can be cleaned mechanically, with a tool known as a pig, or chemically. Mechanical cleanings are the most commonly accepted practice to remove deposits within the pipeline. The pig is passed through the pipeline, scraping away at deposits until it is retrieved in a relatively clean condition. Because operators usually cannot see the inside of the pipeline while cleaning, the cleanliness of the pig was historically used as a proxy for the cleanliness of the pipe. However, assuming the pipeline is clean when the pig is clean can be inaccurate because the mechanical pig is not capable of perfectly removing deposits. As it passes through the pipe, it smears a small amount of debris along the surface, creating a thin coating. This thin layer is compacted over multiple cleanings. It is possible for water and other corrosive elements to be trapped underneath this layer, which increases the risk of internal corrosion. The interior coating can also interfere with in-line inspection tools. The historical failures of mechanical pigging have led to the quick adoption of chemical cleaning. Used in conjunction with a mechanical pig, liquid chemicals can remove more debris in fewer runs. While mechanical cleaning can be performed on both active and inactive pipelines, chemical cleaning requires the pipeline to be temporarily deactivated. The section to be cleaned must be closed off, after which it is be filled with liquid chemicals. The flow rate and the pH of the chemicals are monitored as the pipeline is cleaned—both values increase. Once the flow rate and pH level off, the pipeline is flushed and reactivated. The chemical blends used are proprietary but all follow a few key parameters. Ideal cleaning products have the following properties: wetting, to reduce surface tension of deposits; emulsification, to prevent hydrocarbons from redepositing down the line; detergency, to mobilize hydrocarbon deposits; and dispersion, to keep the deposits in suspension by preventing aggregation of particles.

2.4.4 CORRECTIVE ACTION

Simply, corrective action defines allowing a pipeline to run its full course until it then fails; then once failure has occurred, continuing with a strategy to fix said failure. Pipelines are monitored to ensure that knowledge of when failures will occur don't come as a complete surprise and can be attended to immediately reduce damage costs and timeframes.

This form of maintenance allows for the budget to only be used as it is necessary rather than systematically at regular intervals. Cost-effectiveness as well as knowledge are two major benefits of corrective action. When allowing a pipeline to run its full course whilst being monitored, it can be documented how long the raw pipeline will last under the impact of the substance flowing through it, loads applied to it, as well as external factors which may cause the pipeline to deteriorate. This information can allow engineers to decide whether to redesign the section of pipeline which will be replaced, once failure has occurred, or to replace it with the exact same material if the term served by the pipeline is sufficient. This will also help to predict the life span of the future pipelines designed using same material.

Symbols

a	depth of the equivalent rectangular stress block, (mm)
A	the acid-consuming capability of the wall material
A_s	area of tension reinforcement in length b, (mm^2/m)
b	unit length of pipe, 1000 mm
B_1	crack control coefficient for effect of spacing and number of layers of reinforcement
c	the average rate of corrosion (mm/year)
C_1	crack control coefficient for type of reinforcement
d	distance from compression face to centroid of tension reinforcement, (mm)
d_b	diameter of rebar in inner cage, mm
[DS]	dissolved sulfide concentration (mg/L)
f'_c	design compressive strength of concrete, (MPa)
f_y	design yield strength of reinforcement, (MPa)
F	crack width control factor
F_c	factor for effect of curvature on diagonal tension (shear) strength in curved components
F_d	factor for crack depth effect resulting in increase in diagonal tension (shear) strength with decreasing d
F_N	coefficient for effect of thrust on shear strength
h	overall thickness of member (wall thickness), (mm)
i	coefficient for effect of axial force at service load stress
k	acid reaction factor
J	is pH-dependent factor for proportion of H_2S
w	the width of the stream surface
P'	perimeter of the exposed wall
M_s	service load bending moment acting on length b, (Nmm/m)
M_u	factored moment acting on length b, (Nmm/m)
N_s	axial thrust acting on length b, service load condition (+ when compressive, − when tensile), (N/m)
N_u	factored axial thrust acting on length b, (+ when compressive, − when tensile), (N/m)
s	is the slope of the pipeline

t	elapsed time
u	is the velocity of the stream (m/s)
V_b	basic shear strength of length b at critical section
Φ	the average flux of H_2S to the wall
$\varnothing_f$	strength reduction factor for flexure
ϕ_v	strength reduction factor for shear
Δ	reduction in wall thickness due to corrosion, (mm)
Δ_{max}	maximum permissible reduction in wall thickness (structural resistance or limit), (mm)
a	multiplying constant
b	exponential constant
B'	empirical coefficient of elastic support
B_d	maximum width of trench (mm)
C_L	live load distribution coefficient
D_f	shape factor
D_i	inside diameter (mm)
D_L	deflection lag factor
D_m	mean diameter of pipe (mm)
D_o	outside and inside diameter (mm)
E	modulus of elasticity of pipe (kPa)
E'	soil modulus (kPa)
F_y	tensile strength (MPa)
H	height of backfill (mm)
I	moment of inertia per unit length $(Kg.m^2/m)$
K	bedding factor
L	effective length of pipe (mm)
M_n	flexural capacity (MPa)
P	actual buckling pressure (N/m^2)
P_a	accepted probability of failure
P_s	wheel load (kPa)
P_{cr}	critical buckling pressure (kPa)
P_W	hydrostatic pressure (kPa)
R_w	water buoyancy factor
S_f	safety factor
t	time of exposure (year)
t_c	time of failure (year)
r	radius of pipe (mm)
T_a	allowable wall thrust (MPa)
T_{cr}	critical wall thrust (MPa)
W_c	Marston's load per unit length of pipe (kPa)
W_t	initial wall thickness (mm)
Y	vertical deflection of pipe (mm)
Y_{max}	maximum longitudinal deflection (mm)
y_0	distance from centroid of pipe to the furthest surface of pipe (mm)
Δ	depth of corrosion (mm)
ΔX	ring deflection (mm)
φ	capacity factor
γ_s	unit weight of soil (kN/m^3)

References

Ahrary, A., Nassiraei, A.A.F., Ishikawa, M., 2007. A study of an autonomous mobiles robot for a sewer inspection system. Artif. Life Robotics 11, 23−17.

Dirksen, J., Clemens, F.H.L.R., Korving, H., Cherqui, F., Le Gauffre, P., Snaterse, C., 2013. The consistency of visual sewer inspection data. Struct. Infrastruct. Eng. 9 (3), 214−228.

Duran, O., Althoefer, K., Seneviratne, L.D., 2007. Automated pipe defect detection and categorization using camera/laser-based profiler and artificial neural network. IEEE Trans. Autom. Sci. Eng. 4 (1), 118−126.

Fisk, P., Marshall, J., 2010. Detecting Deteriorating Thinning PCCP Pipe Mortar Coating. Pipelines 920−924.

Gabriel, L.H., 2011. Corrugated Polyethylene Pipe Design Manual and Installation Guide. Plastic Pipe Institute, Irving, TX.

Gorenc, B.E., Syam, A., Tinyou, R., 2005. Steel Designers' Handbook. University of New South Wales Press (UNSW Press), Australia.

Guo, W., Soibelman, L., Garret Jr., J.H., 2009. Automated defect detection for sewer pipeline inspection and condition assessment. Autom. Constr. 18, 587−596.

Hao, T., Rogers, C.D.F., Metje, N., Chapman, D.N., et al., 2012. Condition assessment of the surface and buried infrastructure. Tunnel. Underground Space Technol. 28, 331−334.

Hertzberg, R.W., 1996. Deformation and Fracture Mechanics of Engineering Materials. Wiley, Chichester.

Horoshenkov, K.V., BinAli, M.T., Tait, S.J., 2011. Rapid detection of sewer defects and blockages using acoustic-based instrumentation. Water Sci. Technol. 64 (8), 1700−1707.

Horoshenkov, K.V., Yin, Y.A., Schellart, A., Ashley, R.M., Blanksby, J.R., 2004. The acoustic attenuation and hydraulic roughness in a large section sewer pipe with periodical obstacles. Water Sci. Technol. 50 (11), 97−104.

Koo, D.H., Ariaratnam, S.T., 2006. Innovative method for assessment of underground sewer pipe condition. Autom. Constr. 15, 479−488.

Korving, H., 2004. Probabalistic Assessment of the Performance of Combined Sewer Systems. Delft University of technology, Delft, The Netherlands [Ph.D. thesis].

Korving, H., Clemens, F., 2005a. Impact of dimension uncertainty and model calibration on sewer system assessment. Water Sci. Technol. 52 (5), 35−42.

Korving, H. and Clemens, F. (2005b) 'Reliability of coding of visual inspections of sewers.' *International Conference on Sewer Processes and Networks*, Funchal, Madeira, Portugal.

Laham, S.Al., 1999. Stress Intensity Factor and Limit Load Handbook, Structural Integrity Branch British Energy Generation Ltd, EPD/GEN/REP/0316/98, ISSUE 2.

Lewis, R., Fisk, P., 2005. Detection of prestressed concrete cylinder pipe thinning from hydrogen sulfide deterioration. Pipelines 965−979.

Litvin, I.Y., Alikin, V.N., 2003. Assessment of Reliability Indicators of Main Pipelines. Nedra, Moscow, 2003.

Lohr, K.R., Rose, J.L., 2003. Ultrasonic guided wave and acoustic impact methods for pipe fouling detection. J. Food Eng. 56, 315−324.

Mahmoodian, M., 2013. Reliability Analysis and Service Life Prediction of Pipelines. University of Greenwich, London [Ph.D. thesis].

Moser, A.P., 2010. Buried Pipe Design. McGraw-Hill Professional Pub.

Moser, A.P., Folkman, S., 2008. Buried Pipe Design. McGraw-Hill Professional Pub.

Ritter, M. and Frey. C.W. (2010) 'Rotating optical geometry sensor for inner pipe-surface reconstruction.' Proc. SPIE 7538, Image Processing: Machine Vision Applications III, San Jose, California, January 17, 2010. 7538. 03.

Rogers, C.D.F., Hao, T., Burrow, M.P.N., Costello, S.B., et al., 2012. Condition assessment of the surface and buried infrastructure - a proposal for integration. Tunnel. Underground Space Technol. 28, 202−211.

Sarshar, N., Halfawy, M.R., Hengmeechai, J., 2009. Video processing techniques for assisted CCTV inspection and condition rating of sewers. J. Water Manage. Model NRCC-50451.

Stanić, N., VanDerSchoot, W.P.J., Kuijer, B., Langeveld, J.G., Clemens, F.H.L.R. (2013) 'Potential of laser scanning for assessing structural condition and physical roughness of concrete sewer pipes.' 7th International Conference on Sewer Processes & Networks, SPN7, August 28−30, Sheffield, UK.

Further Reading

Hong, H.P., 1999. Inspection and maintenance planning of pipeline under external corrosion considering generation of new defects. Struct. Saf. 21, 203−222.

Isa, D., Rajkumar, R., 2009. Pipeline defect prediction using support vector machine. Appl. Artif. Intell. 23 (8), 758−771.

Khodayari-Rostamabad, A., Reilly, J.P., Nikolova, N.K., Hare, J.R., Pasha, S., 2009. Machine learning techniques for the analysis of magnetic flux leakage images in pipeline inspection. Electr. Comput. Eng. Dep. 45 (8), 3073−3084.

Chapter 3

Methods for Structural Reliability Analysis

Abstract

This chapter explains the basics of structural reliability analysis which can be used for structural assessment of in-service structures and infrastructures. The aim in structural reliability analysis is calculation of failure probability in which failure is defined as violation of limit state function.

Structural systems and approaches to estimate their reliability, depending on the configuration of the system, will be discussed in Sections 3.4 and 3.5. The evaluation of the contributions of various uncertain parameters associated with pipeline life assessment can be carried out by using sensitivity analysis techniques. For a comprehensive reliability analysis, sensitivity analysis needs to be conducted to indicate variables which affect reliability most. Parametric sensitivity analysis approach including calculation of sensitivity indexes will also be discussed in this chapter.

The basic knowledge of structural reliability analysis is the prerequisite for Chapter 4, Time-Dependent Reliability Analysis, to better understand time-dependent reliability analysis methods for calculation of failure probability of deteriorating pipelines.

Chapter Outline

79

Reliability and Maintainability of In-Service Pipelines. DOI: https://doi.org/10.1016/B978-0-12-813578-5.00003-2

3.1 Background

Reliability analysis and the prediction of the service life of pipelines is one of the major challenges for infrastructure managers and maintenance engineers. Historically, reliability theory has most often been introduced in the military, aerospace, and electronics fields (Cheung and Kyle 1996). Over the past decades, the significance of reliability theory has been increasingly realized in the area of pipeline engineering. The structural reliability began as a subject for academic research about 50 years ago (Freudenthal 1956). The topic has grown rapidly during the last three decades and has evolved from being a topic for academic research to a set of well-developed or developing methodologies with a wide range of practical applications.

Structural reliability can be defined as the probability that the structure under consideration has a sufficient performance throughout its service life. Reliability methods are used to estimate the service life of structures.

In addition to the prediction of initial service life, reliability methods are effective tools to evaluate the efficiency of repair and replacement. The impact of any repair and maintenance option upon the future performance of the structure can be evaluated by decision makers using reliability analysis methods.

Furthermore, reliability analysis of a pipe or a pipeline network can be used at the conceptual design stage to evaluate various design choices and to determine the impact that their implementation could have upon their service lives.

The uncertain nature of the loadings and the performance aspects of pipelines could have led the planners to probabilistic approaches for service life assessment. In probabilistic methods for dealing with uncertainties, the safety and service/performance requirements are measured by their reliabilities. The reliability of a pipeline is defined as its probability of survival (Melchers 1999):

$$P_s = 1 - P_f \qquad (3.1)$$

where

P_s: probability of survival;
P_f: probability of failure.

Failure can be defined in relation to different possible failure modes, commonly referred to as limit states. Reliability is considered to be the probability that these limits will not be exceeded and is equal to the probability of survival. Each of the limit state function variables is attributed to a probability density function that presents its statistical properties.

To summarize, structural reliability analysis can be generally used for the following purposes:

- Service life prediction of in-service pipelines, for funding allocation to most critical parts of the pipeline or network.
- Evaluation of the effect of repair, maintenance, and rehabilitation actions on the service life of the pipeline (ability to examine the consequences of potential action or inaction relative to operational and maintenance procedures).
- To be used at the conceptual design stage to evaluate various design choices and to determine the impact that their implementation would have upon the service lives.

To predict the service life of in-service pipeline, information is required on the present condition of the pipeline, rates of degradation, past and future loading, and definition of the failure of the pipeline. Based on remaining life predictions, cost−benefit analysis can also be made on whether or not a pipeline should be repaired, rehabilitated, or replaced.

3.2 Theory of Reliability Analysis

In the past, the design of pipelines considered all loads and strengths as deterministic values. The strength of a pipeline was determined in such a way that it withstood the load within a certain margin. The ratio between the strength and the load was denoted a safety factor.

This safety factor was considered as a measure of the reliability of the pipeline. However, uncertainties in the loads, strengths, and in the modeling of the pipeline require that methods based on probabilistic techniques to be used. A pipeline is usually required to have a satisfactory performance in the expected service life, i.e., it is required that it does not collapse or become unsafe and that it fulfils certain functional requirements.

In order to estimate the reliability by using probabilistic concepts it is necessary to introduce stochastic variables and/or stochastic processes/fields and to introduce failure and nonfailure behavior of the pipeline under consideration.

Generally the main steps in a reliability analysis for service life prediction are:

1. Identify the significant failure modes of the pipeline.
2. Decompose the failure modes in series systems or parallel systems of single components (only needed if the failure modes consist of more than one component).
3. Formulate failure functions (limit state functions) corresponding to each component in the failure modes.
4. Identify the stochastic variables and the deterministic parameters in the failure functions. Further specify the distribution types and statistical parameters for the stochastic variables and the dependencies between them.
5. Estimate the reliability of each failure mode (illustration of how reliability, or inversely the probability of failure, changes with time).
6. Evaluate the reliability result by performing sensitivity analyses.

Typical failure modes to be considered in a structural reliability analysis of pipeline are yielding, corrosion, buckling (local and global), fatigue, and excessive deformations.

The failure modes (limit states) are generally divided into:

- *Ultimate limit states* correspond to the maximum load-carrying capacity which can be related to, e.g., formation of a mechanism in the pipeline, excessive plasticity, rupture due to corrosion, and instability.
- *Conditional limit states* correspond to the load-carrying capacity if a local part of the pipeline has failed. A local failure can be caused by an accidental action or by fire. The conditional limit states can be related to, e.g., formation of a mechanism in the pipeline, exceedance of the material strength, or instability (buckling).
- *Serviceability limit states* are related to normal use of the pipeline, e.g., excessive deflections, local damage, and excessive vibrations.

The fundamental quantities that characterize the behavior of a pipeline are called the basic variables and can be denoted $X = (X1, \ldots, Xn)$ where n is the number of basic stochastic variables. Typical examples of basic variables are loads, strengths, dimensions, and material properties. The basic variables can be dependent or independent. A stochastic process can be defined as a random function of time in which for any given point in time the value of the stochastic process is a random variable.

The uncertainty modeled by stochastic variables can be divided into the following groups:

- *Physical uncertainty*: (or inherent uncertainty) Is related to the natural randomness of a quantity, for example, the uncertainty in the yield stress due to production variability.
- *Measurement uncertainty*: Is the uncertainty caused by imperfect measurements of, for example, a geometrical quantity.
- *Statistical uncertainty*: Is due to limited sample sizes of observed quantities.
- *Model uncertainty*: Is the uncertainty related to imperfect knowledge or idealizations of the mathematical models used or uncertainty related to the choice of probability distribution types for the stochastic variables.

All the above types of uncertainty can usually be treated by the reliability methods. Another type of uncertainty which is not covered by these methods is gross errors or human errors. These types of errors can be defined as deviation of an event or process from acceptable engineering practice.

Generally, methods to measure the reliability of a pipeline can be divided in four groups, see Madsen et al. (1986):

- Level I methods: The uncertain parameters are modeled by one characteristic value, as, for example, in codes of practice based on the partial safety factor concept.
- Level II methods: The uncertain parameters are modeled by the mean values and the standard deviations, and by the correlation coefficients between the stochastic variables. The stochastic variables are implicitly assumed to be normally distributed. The reliability index method is an example of a level II method.
- Level III methods: The uncertain quantities are modeled by their joint distribution functions. The probability of failure is estimated as a measure of the reliability.
- Level IV methods: In these methods the consequences (cost) of failure are also taken into account and the risk (consequence multiplied by the probability of failure) is used as a measure of the reliability. In this way different designs can be compared on an economic basis taking into account uncertainty, costs, and benefits.

Level I methods can, e.g., be calibrated using level II methods, level II methods can be calibrated using level III methods, etc.

Several techniques can be used to estimate the reliability for levels II and III methods, including the following methods:

Simulation techniques: Samples of the stochastic variables are generated and the relative number of samples corresponding to failure is used to estimate the probability of failure. The simulation techniques are different in the way the samples are generated. Monte Carlo method is the major simulation method for structural reliability analysis of pipelines.

FORM techniques: In First-Order Reliability Methods (FORM) the limit state function (failure function) is linearized and the reliability is estimated using level II or III methods.

SORM techniques: In Second-Order Reliability Methods (SORM) a quadratic approximation to the failure function is determined and the probability of failure for the quadratic failure surface is estimated.

Time-dependent reliability techniques: When a pipeline is subjected to a time-dependent degradation process, probabilistic time-dependent methods can be used. First passage probability theory has been introduced for time-dependent reliability analysis (Melchers 1999). Gamma process concept also has the potential of usage as a model for reliability analysis of pipelines subject to monotonic degradation processes (Mahmoodian and Alani 2014). These methods are discussed and developed in Chapter 4, Time-Dependent Reliability Analysis, for reliability analysis of pipelines.

For a detailed introduction to structural reliability theory references are made to the following textbooks: Melchers (1999), Thoft-Christensen and Baker (1982), and Ditlevsen and Madsen (1996).

3.3 Generalization of a Basic Reliability Problem

In a basic reliability problem only one load effect, S, can be resisted by one resistance, R. The load and the resistance are expressed by a known probability density function, f_S and f_R, respectively.

Considering the definition of safety, the pipeline will be marked as failed if its resistance, R, is less than the stress resultant, S, action on it. Therefore the probability of failure can be stated as follows:

$$P_f = P[R - S \leq 0] = P[G(R,S) \leq 0] \tag{3.2}$$

where $G(R,S)$ is termed the limit state function, and the probability of failure is identical to the probability of limit state violation. In Fig. 3.1, the above equation is represented by the hatched failure domain D, so that the failure probability becomes (Melchers 1999):

$$P_f = P(R - S \leq 0) = \int_D \int f_{RS}(r,s) dr ds \tag{3.3}$$

where $f_{RS}(r,s)$ is the joint (bivariate) density function.

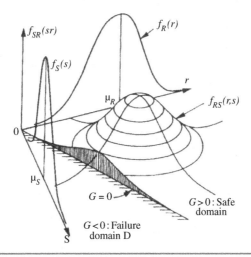

Figure 3.1 Two random variable joint density function $f_{RS}(r, s)$, marginal density functions f_R and f_S and failure domain D (Melchers 1999).

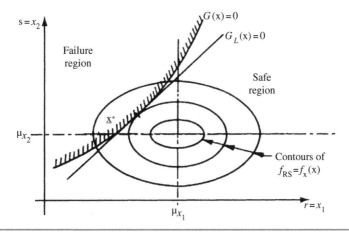

Figure 3.2 Limit state surface $G(X) = 0$ and its linearized version $G_L(X) = 0$ in the space of the basic variables (Melchers 1999).

With the limit state function expressed as $G(X)$, the generalization of Eq. (3.3) becomes:

$$P_f = P[G(X) \leq 0] = \int \cdots \int_{G(X) \leq 0} f_X(X) d_X \qquad (3.4)$$

Here $f_X(X)$ is the joint probability density function for n-dimensional vector X of basic variables. Fig. 3.2 shows a generalization of the reliability problem.

When both the load effect (S) and the pipe resistance (R) are independent and of normal distribution, the integral in Eq. (3.3) can be determined from (Melchers, 1999):

$$P_f(t) = \Phi\left[\frac{-(\mu_R - \mu_S)}{(\sigma_S^2 + \sigma_R^2)^{\frac{1}{2}}}\right] = \Phi(-\beta) \tag{3.5}$$

where Φ is the standard normal distribution function, μ is the mean and σ is the standard deviation of random variables. β is known as safety index or reliability index.

3.4 Reliability of Structural Systems

In some cases of reliability analysis even in a simple pipeline composed of just one segment, various limit states such as bending action, shear, buckling, axial stress, deflection, etc., may apply. Such a composition is referred to as the "structural system."

Individual pipeline components and subsystems have typical service life spans that do not necessarily coincide with one another. In the reliability evaluation of structural systems it is described how the individual limit states interact with each other and how the overall systems reliability can be estimated when the individual failure modes are combined in a series or parallel system.

In a series system (also called a weakest link system), attainment of any one element limit state constitutes failure of the pipeline. All components of a parallel system (also called a redundant system) must fail for a system failure to occur. Combining parallel and series subsystems can make more complex systems (Fig. 3.3).

If in a system reliability problem each failure mode is represented by a limit state equation $G_i(X) = 0$ in basic variable space, the direct extension of the fundamental reliability problem (Eq. (3.2)) is

$$P_f = \int_{D \in X} \ldots \int f_X(X) d_X \tag{3.6}$$

where X represents the vector of all basic random variables (load, strength of members, member properties, sizes, etc.) and D (and $D1$) is the domain in X defining failure of the system. This is defined in terms of the various failure modes as $G_i(X) \leq 0$.

In Fig. 3.4 expression of Eq. (3.6) is defined in a two-dimensional X space.

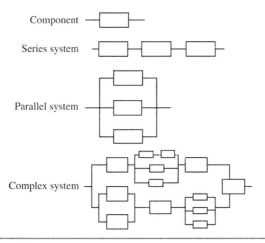

Figure 3.3   System definitions.

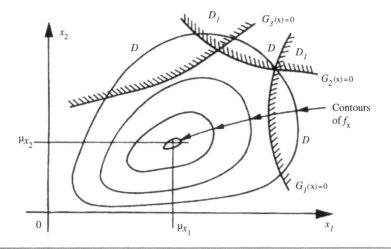

Figure 3.4   Basic structural system reliability problem in two dimensions showing failure domain D (and D1) (Melchers, 1999).

As it was defined previously, a parallel system can fail only when all its contributory components have reached their limit states. This means that, in contrast with the situation for series systems, the behavioral characteristics of the system components are significantly important in defining system failure.

For a pipe, the occurrence of failure due to violation of each limit state function will constitute its total failure. Therefore, a series system is more suitable for the failure assessment of pipes. According to the theory of systems reliability the failure probability for a series system at time t $(P_{f,s}(t))$ can be estimated by (Thoft-Christensen and Baker, 1982):

$$\max[P_{f,i}(t)] \leq P_{f,s}(t) \leq 1 - \prod_{i=1}^{n}[1 - P_{f,i}(t)] \qquad (3.7)$$

where $P_{f,i}(t)$ is the failure probability of the pipe due to the ith failure mode at time t and n is the number of failure modes.

3.5 Sensitivity Analysis

Sensitivity analysis is widely accepted as a necessary part of reliability analysis of structures and infrastructure. The effect of variables on the reliability of a pipeline can be analyzed by doing a comprehensive sensitivity analysis. In view of the large number of variables that affect the limit state function, it is of interest to identify those variables that affect the failure most so that more research can focus on those variables.

Sensitivity analysis should be carried out to provide quantitative information necessary for classifying random variables according to their importance. These measures are essential for reliability-based service life prediction of deteriorating pipelines.

Sensitivity analysis provides the degree of variation of limit state functions or measures at a specific point characterized by a realization of all random variables. Similarly to the conventional sensitivity measure in the reliability approaches, the sensitivity measure, S, can be defined as follows (Kong and Frangopol, 2005):

$$S_{G(X)}(X_i) = \frac{\partial G(X)}{\partial X_i} = \lim_{\varepsilon \to 0} \frac{G(X + \varepsilon) - G(X)}{\varepsilon} \qquad (3.8)$$

where G is a performance function of X; X and ε are vectors; and ε is a small perturbation. An element X_i of X can be any type of variable or parameter. For instance, it can be a mean or a standard deviation of a variable, or a deterministic parameter. For a complex system, the sensitivity measure can be computed by using the numerical differentiation method rather than by an analytical approach (Kong and Frangopol, 2005).

Different sensitivity indexes have been introduced. In this section, relative contribution, sensitivity ratio (SR), and omission factor are discussed.

3.5.1 RELATIVE CONTRIBUTION

A sensitivity index that can be used in a comprehensive reliability analysis is the relative contribution of each variable in limit state function. The relative contribution (α_x^2) of each random variable (x) to the variance of the limit state function is introduced as follows (Ahammed and Melchers, 1994):

$$\alpha_x^2 = \frac{\left(\frac{\partial_G}{\partial_x}\sigma_x\right)^2}{\sigma_G^2} \tag{3.9}$$

where σ_x is standard deviation of the random variable x and σ_G^2 is the variance of the limit state function. Variables with higher values of α_x^2 contribute more in limit state function than other variables; therefore more focus and study needs to be carried out to determine the accurate values for such variables.

3.5.2 SENSITIVITY RATIO

A method of sensitivity analysis applied in many different models in science, engineering, and economics is the SR, also known as the elasticity equation. The ratio is equal to the percentage change in output (e.g., probability of failure) divided by the percentage change in input for a specific input variable, as shown in the following equation (EPA 540, 2001):

$$SR = \frac{\left(\frac{Y_2 - Y_1}{Y_1}\right) \times 100\%}{\left(\frac{X_2 - X_1}{X_1}\right) \times 100\%} \tag{3.10}$$

where Y_1 = the baseline value of the output variable using baseline values of input variables;

Y_2 = the value of the output variable after changing the value of one input variable;

X_1 = the baseline point estimate for an input variable;

X_2 = the value of the input variable after changing X_1.

Risk estimates are considered most sensitive to input variables that yield the highest absolute value for SR. The basis for this equation can be understood by examining the fundamental concepts associated with partial derivatives. In fact, SR is equivalent to the normalized partial derivative. Variables with higher values of sensitivity ratios are more effective on the limit state function or the probability of failure (EPA 540, 2001).

3.5.3 OMISSION SENSITIVITY FACTOR

In computing the reliability index, the basic random variables X can be transformed into standardized normal space U and the limit state function, $G(X,t)$, can be transformed to $g(U,t)$ (Melchers, 1999).

The effect of randomness of variables on the probability of failure can be measured by an omission sensitivity factor. According to Ditlevsen and Madsen (1996), the omission sensitivity factor with respect to random variable u_i can be determined by:

$$\zeta_{u_i}(t) = \frac{\beta(t)|_{U_{i(t)=u_i}}}{\beta(t)} = \frac{1 - \alpha_i u_i / \beta(t)}{\sqrt{1 - \alpha_i^2}} \qquad (3.11)$$

where α is the normal unit vector to the limit state surface $g(U, t)$ at checking point u^* and time t (Melchers, 1999). As can be seen the omission sensitivity factor measures the relative error in the value of reliability index β if an input random variable is replaced by a fixed value (i.e., treated as a deterministic variable). Thus when the relative error of random variables (i.e., omission sensitivity factor) is around one ($\zeta_{u_i} \approx 1$), it may be appropriate to treat them as deterministic variables if the full statistical information of them is not available.

3.6 Background and Methods for Reliability Analysis of Pipes

Since large investment is required for building new pipeline networks for different purposes such as urban water supply and/or energy infrastructure, it is unlikely to replace the existing pipeline networks completely over a short period of time. Therefore, the resort has to be maintenance and rehabilitation of existing pipelines. To have an optimum strategy for maintenance and rehabilitation plans in the management of a pipeline asset, accurate prediction of the service life of in-service pipelines is essential. But this cannot be achieved without an accurate method for reliability analysis in which the likelihood of pipeline failure is determined.

Reliability analysis can cover a wide domain of failure assessment of structures and infrastructure including both service life prediction and failure rate prediction. It should be noted that there is a clear distinction between the two terms:

Failure rate prediction of pipes: When the result of reliability analysis and/or failure assessment is presented as a number of failures within a period of time (e.g., breaks/year), it should be ideally considered as failure rate prediction.
Service life prediction of pipes: When in a study, service life of pipe(s), in terms of time, is investigated; the study should be named as service life prediction.

As a comparison, there is considerably less literature in the field of service life prediction compared to the failure rate prediction of pipes. It needs to be clarified that the focus of this book is on service life prediction of pipelines. To that effect, the existing literature on reliability analysis and service life prediction of energy pipelines concrete sewers and ferrous pipes (including cast iron (CI) water pipes) is presented in this section.

Kleiner and Rajani (2001) defined two main classes of methods for service life prediction: deterministic and probabilistic methods. Deterministic methods do not consider variation in any variables that affect pipe behavior and failure, whilst probabilistic methods consider some or all variables as random variables. In another classification Clair and Sinha (2012) classified deterioration models available for predicting service life of water pipes into six categories: deterministic, statistical, probabilistic, artificial neural networks (ANN), fuzzy logic, and heuristic. The input parameters and output results of pipe deterioration models are heavily dependent on the type of methodology chosen. To find out the gaps and limitations of each model, the models are briefly explained here.

3.6.1 DETERMINISTIC MODELS

Deterministic models often use laboratory tests and sample specimens to find the necessary information, therefore the relationships between components are certain. Variations and uncertainties in variables are not considered in deterministic methods, while probabilistic methods consider some or all variables as random variables.

Kaempfer and Berndt (1999) undertook a laboratory experiment to predict service life of concrete sewers subjected to sulfide corrosion. They used deterministic parameters from an accelerated laboratory test to predict the service life of concrete sewers. Since their study was in laboratory conditions, they suggested that for a more accurate result the data from real sewage conditions are necessary.

Rajani et al. (2000) proposed a method to estimate the remaining service life of CI water pipes by considering that the corrosion pits reduce the structural capacity of the pipes. The residual capacity of the pipes was calculated by a reiterative model, based on corrosion pit measurement and the anticipated corrosion rate. The method was deterministic in the view that it does not consider the uncertainties involved in all factors contributing to the corrosion and subsequent failures.

A comprehensive deterministic life time assessment has been carried out by Kienow and Kienow (2001). They performed sulfide corrosion modeling as part of screening analyses to support the prioritization of sewer evaluation efforts for a sewer inspection and evaluation in the City of Fresno, CA. The project consisted of the inspection and evaluation of approximately 90 km of concrete sewers in sizes ranging from 30 to 70 cm in diameter.

Deb et al. (2002) presented a deterministic model based on analyzing the growth of corrosion pits on CI pipes, loss of wall thickness, and the strength reduction of the pipe over time. Kim et al. (2007) developed prediction models for CI pipes using the assessment of residual tensile strength based on pit characteristics and fracture toughness. Results illustrated that the proposed models using tensile strength and fracture toughness of CI pipes successfully estimated the residual life of water pipes. Analysis of the results showed that the determination of fracture toughness may be more reliable than considering only the pit depth.

3.6.2 Probabilistic Modes

Probabilistic models are often used when historical failure or inspection data is limited or unavailable. These models specifically analyze the effective parameters on pipe performance rather than evaluating the previous pipe failure history. Uncertainties are involved by considering random variables. Usually, these methods are applied to pipes where the process of deterioration and factors for failure are well understood.

Various frameworks have been proposed to model the behavior of underground pipelines for different types of material, using the reliability-based concept. Ahammed and Melchers (1994, 1995, 1997) reported a comprehensive and continuous study on the reliability analysis of underground steel pipelines. To consider uncertainty associated with the rate of corrosion and the uncertain location of its occurrence, they used a probabilistic approach (first-order second-moment reliability method, FOSM) for the analysis of pipeline reliability.

In 1994, they defined the failure mode as exceeding the sum of total stresses from the maximum allowable stress (yield strength of the pipe). They considered three types of existing stresses caused by internal pressure and external pressure:

- The circumferential stress due to internal fluid pressure.
- The bending stress in the circumferential direction produced in the pipe wall by external soil loading.
- The circumferential bending stresses produced in the pipe wall due to external traffic loads.

Taking into account an empirical time-dependent corrosion model, resulted in a nonlinear limit state function that required an iterative solution technique for the calculation of reliability index and for sensitivity analysis.

In 1995 they modeled the growth of corrosion pits to assess the service life of liquid carrying metallic pipelines. In the pitting (localized corrosion) model, two corrosion related parameters (including pipeline dimension and liquid flow) were treated as probabilistic variables.

The limit state function was defined as the difference between the allowable fluid loss and the estimated fluid loss through the pit holes.

To calculate the probability of failure, they used the level II FOSM reliability method. This method requires an iterative solution, when the random variables are not normally distributed and the limit state function is nonlinear.

They developed their model in further studies (Ahammed and Melchers, 1997) by considering the effect of following longitudinal stresses:

- Longitudinal tensile stresses as a result of Poisson's ratio effect from the outward radial action of the internal fluid pressure.
- Longitudinal thermal stresses.
- Longitudinal stresses due to bending as a result of unevenness or settlement of the pipeline bedding.

Camarinopoulos et al. (1999) used a combination of approximate quadrature analytical and Monte Carlo method to evaluate the multiple integrals in their reliability analysis for cast iron buried water pipes. They also used the model to assess the sensitivity of structural reliability to the variation of some important parameters such as wall thickness, unsupported length, and external corrosion coefficient.

Yves and Patrick (2000) also presented a method to calculate the reliability of the buried water pipes using maintenance records and the Weibull distribution for underlying variables. The method appears to rely entirely on the historical data, which in most cases is unknown.

Benmansour and Mrabet (2002) studied the reliability of buried concrete sewers by using the FOSM method. They studied two typical cases to assess the effect of loads on the circumferential and longitudinal behavior of pipes. Therefore two limit state functions that they considered in their study were based on longitudinal cracking and circular cracking due to bending moments. They did not consider corrosion as a deterioration process, therefore their methodology was a simple time-independent method.

Sinha and Pandey (2002) developed a model to estimate the failure probability of aging pipelines prone to corrosion by using simulation-based probabilistic neural network analysis. The approximations in their neural network model can be considered as a limitation of their methodology.

Sadiq et al. (2004) used Monte Carlo simulations to perform the reliability analysis of CI water mains, considering axial and hoop stresses as acting loads in a limit state function. The reduction in the factor of safety (FOS) of water mains over time was computed, with a failure defined as a situation in which FOS becomes smaller than one. The Monte Carlo simulations yielded an empirical probability density function of time to failure, to which a lognormal distribution was fitted leading to the derivation of a failure hazard function.

Davis et al. (2005) used Weibull probability distribution to account for variation in the degradation rate of asbestos cement sewers. A tensile failure model was developed that simulates degradation until failure under in-service loading conditions. Simulated failure times were then fitted to a Weibull probability distribution, which allows the expected time to first failure to be calculated at different locations along the pipeline. They found a reasonable agreement between predicted failure times and recorded failures for a period of 8 years.

Amirat et al. (2006) used reliability analysis to assess the effect of both the residual stresses generated during manufacturing process and in-service corrosion of underground steel pipes. During the service life of a pipe, residual stress relaxation occurs due to the loss of pipe thickness as material layers are consumed by corrosion. First they focused on the influence of residual stresses in uncorroded pipelines in order to identify the sensitivity of system parameters. In the second step, a probabilistic-mechanical model was used to couple the residual stress model with the corrosion model, in order to assess the aging effects through the pipe service life. For long-term corrosion, the reliability analysis incorporated the residual stress relaxation resulting from wall thickness losses. The probability of failure of the pipeline was then evaluated for different corrosion rates varying from the atmospheric baseline to very active corrosion processes.

DeSilva et al. (2006) presented a condition assessment and probabilistic analysis to estimate failure rates in metallic pipelines. A level II FOSM analysis was combined with condition assessment data to determine the probability of failure. Davis and Marlow (2008) developed a physical probabilistic failure model for service life prediction of CI pipelines subject to corrosion under internal pressure and external loading. A limitation within their study was that the model only considered internal pressure and in-plane bending; therefore, the resulting failure mode was only shown with a longitudinal fracture, which is just one type of several failure modes. Other failure modes, such as circumferential fractures, were not considered.

Moglia et al. (2008) looked at the exploration of a CI pipe failure model utilizing fracture mechanics of the pipe failure process. The first model generated was simple, which allows explorations of additional model assumptions. Throughout numerous assumptions, the model improved drastically. An elementary method, FOSM, was initially used but proved to yield inaccurate results. A new approach to the model evaluated the nominal tensile strength of pipes, which could determine the maximum corrosion defect. To account for the uncertainty or randomness within the data, a Weibull distribution was utilized adding stochasticity to the corrosion rate. The proposed model calculated failure rates based on historical data using a random Poisson statistical process. The maximum likelihood estimator used within the Poisson distribution was used to calculate the failure rate of the historical data sets. A case study was employed utilizing small diameter reticulation mains. By modeling various assumptions into the simulated model, the predicted and observed failure rates yielded similar results. Only failure modes by corrosion or combined corrosion and fractures were included within the observed data model.

Teixeira et al. (2008) used both Monte Carlo simulation method and FORM for assessing the reliability of corrosion affected pipelines subjected to internal pressure.

A methodology for predicting remaining life of corroded underground steel gas pipelines was presented by Li et al. (2009). They took the effect of randomness of pipeline corrosion into account by developing a mechanically-based probabilistic model. Monte Carlo simulation was employed in their study to calculate the remaining safe life and its cumulative distribution function.

Yamini (2009) also used different reliability analysis methods (Monte Carlo simulation, FORM, and SORM) for failure analysis of CI water mains. In his study two failure modes were considered individually. A failure mode was defined as the point at which the corrosion depth is more than the maximum acceptable decrease in pipe wall thickness and another failure mode was defined as the time at which total stresses exceed the pipe strength capacity.

Lee et al. (2010) also used a FORM to evaluate the time-dependent reliability index for a fully deteriorated piping component rehabilitated with fiber-reinforced plastic (FRP), considering the demand of internal fluid pressure, external soil pressure, and traffic loading.

Zhou (2011) developed a methodology to carry out the time-dependent reliability evaluation of a pressurized steel gas pipeline containing an active corrosion defect by taking into account the time-dependency of the internal pressure.

A methodology for predicting remaining life of corroded underground steel gas pipelines presented by Li et al. (2009). They took the effect of randomness of pipeline corrosion into account by developing a mechanically-based probabilistic model. Monte Carlo simulation was employed in their study to calculate the remaining safe life and its cumulative distribution function. Li and Mahmoodian (2013) used an analytical first passage reliability method for estimation of failure probability of corrosion affected CI water pipes. Their study was limited to uniform corrosion rate; while in practice aging pipelines normally possess corrosion pits or cracks.

Qin (2014) developed a Monte Carlo simulation-based methodology to evaluate the time-dependent system reliability of corroding pipelines in terms of three different potential failure modes, namely small leak, large leak, and rupture. In his study only the corrosion depth was considered as the time-dependent parameter and corrosion length was treated as a time-independent parameter.

Mahmoodian and Li (2015) developed a stochastic model for the stress intensity factor and a time-variant analysis method based on gamma process concept to quantify the failure probability. In their study two types of stresses (i.e., hoop and axial) were considered for two cases of corrosion (i.e., external and internal).

A reliability-based methodology for assessment of corroded steel gas pipelines was presented by Mahmoodian and Li (2017). They developed a stochastic model for strength loss which relates to key factors that affect the residual strength of a corroded pipe. The failure of pipeline was defined as when pipe residual strength falls below its operating pressure. An analytical time-variant method was employed to quantify the probability of failure due to corrosion so that the time for the pipeline to fail and hence require repairs, could be determined with confidence. To deal with the assessment of pipelines with more than one corrosion pit, they employed a system reliability analysis method. Monte Carlo simulation technique was applied to verify the results of the analytical method.

3.6.3 OTHER MODELS

Statistical models initially forecast the number of pipe failures with the use of maintenance records and failure data. Recently, some researchers have also used statistical models to predict pipe failure. Kleiner and Rajani (2008) examined the use of a nonhomogenous Poisson model and evaluated factors that affect water pipes. Berardi et al. (2008) proposed the application of performance indicators to model pipe deterioration rate using the method of evolutionary polynomial regression (EPR). Wang et al. (2009) developed deterioration models in their statistical analysis to predict the annual break rates of water mains considering the pipe material type diameter, age, and length.

Compared with deterministic models, statistical models take into account historical data and their variables in the prediction of pipe failures. Statistical models can be applied to pipes that have an adequate and reliable historical database over time. Therefore, applicability of statistical models is limited when considering cases with insufficient monitoring data.

Artificial Neural Networks (ANN) predict pipe deterioration rates by utilizing all variables that influence the service life of a pipe. Each analyzed parameter can increase the system performance and reliability. Parameters can be prioritized by applying weights and learning algorithms. A high level of skill is involved in developing these complex networks. Data preprocessing, training, and testing methods for selecting appropriate network are also required.

Christodoulou et al. (2004) examined the deterioration of water mains using ANN techniques from parametric and nonparametric analyses. Al-Barqawi and Zayed (2008) developed a condition rating model to assess the rehabilitation priority for water pipes using an ANN. The output variable consists of a condition rating scale from 0 to 10 with 0 representing a critical condition and 10 representing an excellent condition.

Compared with other models, ANN modeling necessitates the use of all variables that influence the failure of a pipe. Higher degree of nonlinearity can be taken into consideration by ANN models. Because these models basically depend on actual data parameters, a limitation may be the lack of data utilities possess. An increased level of skill and training is also required in order to develop the complex networks in ANN models (Clair and Sinha, 2012).

Fuzzy logic models use engineering judgment and professional experience in order to predict pipes' deterioration processes. This type of model is often used when data is scarce and observations and model criteria are expressed in vague or "fuzzy" terms. This technique implements expert opinions. A high level of skill is required in constructing the rule set and deciding the defuzzification process for a reliable output.

Kleiner et al. (2005) proposed a fuzzy Markov deterioration process to predict the future condition of CI pipes. The fuzzy set techniques help to incorporate the imprecision and subjectivity of the data. Rajani and Tesfamariam (2007) applied a fuzzy set theory for consideration of uncertainties to estimate the structural capacity of aging CI water mains. Fares and Zayed (2010) developed a hierarchical fuzzy expert system to determine the risk-of-failure of water mains. The system consists of 16 risk-of-failure factors within four main categories: environmental, physical, operational, and post failure.

Compared with other deterioration models, fuzzy logic models necessitate the use of fuzzy logic-based techniques that possess the ability to incorporate engineering judgment to predict pipe deterioration. Fuzzy logic models are used for systems that are subject to uncertainties, ambiguities, and contradictions. The primary limitation for fuzzy logic models is the challenges that exist in constructing a fuzzy rule set, selecting a membership function, and determining a defuzzification process (Clair and Sinha, 2012).

Heuristic models are rare and limited in nature, but can illustrate how methodologies incorporate engineering knowledge in the determination of deterioration rates. This technique is a structured way of capturing expert opinions. A limitation of using engineering knowledge is the inconsistency in the expert judgments from individual to individual and/or lack of personal experience in making the judgments. However, model capabilities can be improved by considering additional expert knowledge and opinions (Clair and Sinha, 2012).

Watson (2004) proposed a Bayesian methodology that combines engineering knowledge with recorded failure data to establish failure rates. This hierarchical model works to combine information from various sources of data. Al-Barqawi and Zayed (2008) presented an integrated model utilizing an analytic hierarchy process (AHP) and ANN. The AHP is first utilized to assign weights and assess the current condition of a water main based on physical, environmental, and operational factors. Then an ANN, which is trained with the available data set, accounts for missing points utilizing pattern recognition. Zhou et al. (2009) developed a pipe condition ranking method using a heuristic outranking method that constructs an outranking relation for a particular criterion and uses this relation to give ranks to each pipe.

A main difference between the heuristic models and other deterioration models is incorporating engineering knowledge rather than data parameters that affect a pipe. This procedure can be considered as a reliable method to illustrate failure risks with limited or no pipe data. However, application of this methodology is limited due to its simplicity and the fact that any type of pipe material can be analyzed. In general, heuristic models can be used as a first step in the determination of failure rates if no other mathematical models are available (Clair and Sinha 2012).

Table 3.1 presents a brief discussion and comparison of all six of the above-mentioned models. As a summary, it can be seen from the above discussion and comparisons that of all these models and methods, one significant feature of pipe failure has not been considered explicitly in one single method in full. A review of most recent research literature (Sadiq et al. 2004; Moglia et al. 2008; Yamini 2009; Clair and Sinha 2012) also suggests that in most reliability analyses for buried pipes, multifailure modes are rarely considered; while the real condition in practice, necessitates consideration of multifailure modes analysis.

TABLE 3.1 Comparison of Pipe Deterioration Models

Model	When to be Applied	Input Variables	Advantages of the Model	Disadvantages of the Model
Deterministic models	The failure modes and mechanisms of the pipe deterioration is well understood	Deterministic parameters taken from laboratory tests or experiment	Predict an average single value of a dependent variable	Applicability of each individual model is restricted to a specific location
Probabilistic models	Historical failure or inspection data is limited or unavailable	Random variables which affect the pipe performance	Entails the prediction for databases that have very little information	Failure mechanism should be well understood and deterioration formulation should be available
Statistical models	Sufficient historical failure or condition data are available	Pipe infrastructure data, Condition of pipe ranking	Applicability to be used for all types of pipe materials	Applicability is limited when considering newer pipes or pipes with an insufficient historical database, not suitable for modeling the actual deterioration process of pipe

(Continued)

TABLE 3.1 (Continued)

Model	When to be Applied	Input Variables	Advantages of the Model	Disadvantages of the Model
ANN models	Deterioration under very complex circumstances and under uncertain environments	All variables that influence the failure of a pipe and the way the nodes and the interconnections are arranged within the layers (topology)	A useful tool for modeling failure rate with high degree of nonlinearity, can be utilized for any type of pipe material	High levels of skill and training are required to develop the complex networks, quality labeled data are required for supervised training
Fuzzy logic models	Data is scarce and observations and model criteria are subject to uncertainties, ambiguities, and contradictions	Variables are assigned a degree of membership on a continuous interval (0, 1) based on expert opinions	incorporate engineering judgment to predict pipe deterioration	Challenges exist in constructing fuzzy rule sets, selecting membership functions, and determining the defuzzification process
Heuristic models	Infrastructure problems are not well understood (limited or no pipe data)	Engineering knowledge	Illustrate failure risks with limited or no pipe data	Application is limited due to its simplicity

References

Ahammed, M., Melchers, R.E., 1994. Reliability of underground pipelines subject to corrosion. J. Transp. Eng. 120 (6), November/December.

Ahammed, M., Melchers, R.E., 1995. Probabilistic analysis of pipelines subjected to pitting corrosion leaks. Eng. Struct. 17 (2).

Ahammed, M., Melchers, R.E., 1997. Probabilistic analysis of underground pipelines subject to combined stress and corrosion. Eng. Struct. 19 (12), 988–994.

Al-Barqawi, H., Zayed, T., 2008. Infrastructure management: Integrated AHP/ANN model to evaluate municipal water mains' performance. J. Infrastruct. Syst. 14 (4), 305–318.

Amirat, A., Mohamed-Chateauneuf, A., Chaoui, K., 2006. Reliability assessment of underground pipelines under the combined effect of active corrosion and residual stress. Int. J. Pres. Ves. Pip. 83, 107–117.

Benmansour, A., Mrabet, Z., 2002. Reliability of Buried pipes, Asranet (Integrating Advanced Structural Analysis with Structural Reliability Analysis) International Colloquium 8–10 July, Glasgow, Scotland.

Berardi, L., Giustolisi, O., Kapelan, Z., Savic, D.A., 2008. Development of pipe deterioration models for water distribution systems using EPR. J. Hydroinform. 10 (2), 113–126.

Camarinopoulos, L., Chatzoulis, A., Frontistou-Yannas, S., Kallidromitis, V., 1999. Assessment of the time-dependent structural reliability of buried water mains. Reliab. Eng. Syst. Safe. 65, 41–53.

Cheung, M.S., Kyle, B.R., 1996. Service life prediction of concrete structures by reliability analysis. Construct. Build. Mater. 10 (1), 45–55.

Christodoulou, S., Aslani, P., Vanreterghem, A., 2004. A risk analysis framework for evaluating structural degradation of water mains in urban settings, using neurofuzzy systems and statistical modeling techniques. Proceedings of the World Water and Environmental Resources Congress and Related Symposia. ASCE, Philadelphia, PA.

Clair St., A.M., Sinha, S., 2012. State-of-the-technology review on water pipe condition, deterioration and failure rate prediction models. Urban Water J. 9 (2), 85−112.

Davis, P., De Silva, D., Gould, S., Burn, S., 2005. Condition assessment and failure prediction for asbestos cement sewer mains, Pipes Wagga Wagga Conference, Charles Sturt University, Wagga Wagga, New South Wales, Australia.

Davis, P., Marlow, D., 2008. Asset management: quantifying economic lifetime of large-diameter pipelines. J. Am. Water Works Assoc. 110−119.

Deb, A.K., Grablutz, F.M., Hasit, Y.J., Snyder, J.K., 2002. Prioritizing Water Main Replacement and Rehabilitation. American Water Works Association Research Foundation, Denver, CO.

De Silva, D., Moglia, M., Davis, P., Burn, S., 2006. Condition assessment and probabilistic analysis to estimate failure rates in buried metallic pipelines. J. Water Supply: Res. Technol.-Aqua 55 (3), 179−191.

Ditlevsen, O., Madsen, H.O., 1996. Structural Reliability Methods. John Wiley and Sons, Chichester.

EPA 540-R-02-002, 2001. Risk Assessment Guidance for Superfund: Volume III − Part A, Process for Conducting Probabilistic Risk Assessment, Office of Emergency and Remedial Response, U.S. Environmental Protection Agency, Washington, DC 20460.

Fares, H., Zayed, T., 2010. Hierarchical fuzzy expert system for risk of failure of water mains. J. Pipeline Syst. Eng. Pract. 1 (1), 53−62.

Freudenthal, A.M., 1956. Safety and the probability of structural failure. Transactions 121, 1337−1397. ASCE.

Kaempfer, W., Berndt, M., 1999. Estimation of service life of concrete pipes in sewer networks. In: Lacasse, M.A., Vanier, D.J. (Eds.), Durability of Building Materials and Components 8. Institute for Research in Construction, Ottawa ON, K1A 0R6, Canada, pp. 36−45.

Kim, J., Bae, C., Woo, H., Kim, J., Hong, S., 2007. Assessment of residual tensile strength on cast iron pipes. In: Proceedings of the Pipelines: Advances and Experiences with Trenchless Pipeline Projects, pp. 1−7.

Kleiner, Y., Rajani, B., 2001. Comprehensive review of structural deterioration of water mains: statistical models. Urban Water 3, 131−150.

Kleiner, Y., Rajani, B., Sadiq, R., 2005. Risk management of large-diameter water transmission mains. American Water Works Association Research Foundation, Denver, CO.

Kleiner, Y., Rajani, B., 2008. Prioritising individual water mains for renewal. Proceedings of the ASCE/EWRI World Environmental and Water Resources. National Research Council Canada-CNRC-NRC, Honolulu, Hawaii, pp. 1−10.

Kong, J.S., Frangopol, D.M., 2005. Sensitivity analysis in reliability-based lifetime performance prediction using simulation. J. Mater. Civil Eng. 17 (3), June 1.

Lee, L.S., Estrada, H., Baumert, M., 2010. Time-dependent reliability analysis of FRP rehabilitated pipes. J. Compos. Construct., ASCE 14 (3), June 1.

Li, C.Q., Mahmoodian, M., 2013. Risk based service life prediction of underground cast iron pipes subjected to corrosion. J. Reliab. Eng. Syst. Safe. 119, 102−108. November.

Li, S.X., Yu, S.R., Zeng, H.L., Li, J.H., Liang, R., 2009. Predicting corrosion remaining life of underground pipelines with a mechanically-based probabilistic model. J. Petrol. Sci. Eng. 65 (3−4), 162−166.

Madsen, H.O., Krenk, S., Lind, N.C., 1986. Methods of Structural Safety. Prentice-Hall, NJ.

Mahmoodian, M., Li, C.Q., 2015. Structural integrity of corrosion-affected cast iron water pipes using a reliability-based stochastic analysis method. Struct. Infrastruct. Eng. 12 (10), 1356−1363. Available from: https://doi.org/10.1080/15732479.2015.1117114.

Mahmoodian, M., Alani, M., 2014. A gamma distributed degradation rate (GDDR) model for time dependent structural reliability analysis of concrete pipes subject to sulphide corrosion. Int. J. Reliab. Saf. 8 (1), 19–32.

Mahmoodian, M., Li, C.Q., 2017. Failure assessment and safe life prediction of corroded oil and gas pipelines. J. Petrol. Sci. Eng. 151, 434–438.

Melchers, R.E., 1999. Structural Reliability Analysis and Prediction, second ed. John Wiley and Sons, Chichester.

Moglia, M., Davis, P., Burn, S., 2008. Strong exploration of a cast iron pipe failure model. Reliab. Eng. Syst. Safe. 93, 863–874.

Qin, H., 2014. Probabilistic Modeling and Bayesian Inference of Metal-Loss Corrosion with Application in Reliability Analysis for Energy Pipelines (Ph.D. dissertation). The University of Western Ontario.

Rajani, B., Makar, J., McDonald, S., Zhan, C., Kuraoka, S., Jen, C.K., et al., 2000. Investigation of Grey Cast Iron Water Mains to Develop a Methodology for Estimating Service Life. American Water Works Association Research Foundation, Denver, CO.

Rajani, B., Tesfamariam, S., 2007. Estimating time to failure of cast iron water mains. Water Manage. 160 (WM2), 83–88.

Sadiq, R., Rajani, B., Kleiner, Y., 2004. Probabilistic risk analysis of corrosion associated failures in cast iron water mains. Reliab. Eng. Syst. Safe. 86 (1), 1–10.

Sinha, S.K., Pandey, M.D., 2002. Probabilistic neural network for reliability assessment of oil and gas pipelines. Comput.-Aided Civil Infrastruct. Eng. 17, 320–329.

Teixeira, A.P., Soares, C.G., Netto, T.A., Estefen, S.F., 2008. Reliability of pipelines with corrosion defects. Int. J. Pressure Vessels Piping 85 (2008), 228–237.

Thoft-Christensen, P., Baker, M.J., 1982. Structural Reliability Theory and Its Applications. Springer-Verlag Berlin, Heidelberg.

Yamini, H., 2009. Probability of Failure Analysis and Condition Assessment of Cast Iron Pipes Due to Internal and External Corrosion in Water Distribution Systems (Ph.D. dissertation). University of British Colombia.

Yves, L.G., Patrick, E., 2000. Using maintenance records to forecast failures in water networks. Urban Water 2, 173–181.

Wang, Y., Zayed, T., Moselhi, O., 2009. Prediction models for annual break rates of water mains. J. Perform. Construct. Fac. 23 (1), 47–54.

Watson, T.G., 2004. Bayesian-based pipe failure model. J. Hydroinform. 6, 259–264.

Zhou, W., 2011. Reliability evaluation of corroding pipelines considering multiple failure modes and time-dependent internal pressure. J. Infrastruct. Syst. 17 (4), 216–224.

Zhou, Y., Vairavamoorthy, K., Grimshaw, F., 2009. Development of a fuzzy based pipe condition assessment model using PROMETHEE. In: Proceedings of the World Environmental and Water Resources Congress: Great Rivers, pp. 1–10.

Further Reading

Kienow, K.E., Kienow, K.K., 2004. Risk management ... predicting your next concrete pipe sewer failure before it happens. ASCE International Conference on Pipelines, San Diego, CA, United States.

van Noortwijk, J.M., Klatter, H.E., 1999. Optimal inspection decisions for the block mats of the Eastern-Scheldt barrier. Reliab. Eng. Syst. Safe. 65 (3), 203–211.

Time-Dependent Reliability Analysis

Chapter

4

Abstract

All in-service pipelines encounter deterioration due to environmental effects. The deterioration is a time-dependent process which makes the structural integrity of pipelines a time variant characteristic. Advanced mathematical methods which employ probability theory need to be used for time-dependent reliability analysis of pipelines. In this chapter, the methods for the time-dependent reliability analysis of in-service pipelines are presented. These methods will allow for:

- Application to all type of corrosion-affected pipelines.
- The capability to consider multi failure mechanisms/modes.
- Consideration of the scarcity of monitoring data from real-world examples.

103

Reliability and Maintainability of In-Service Pipelines. DOI: https://doi.org/10.1016/B978-0-12-813578-5.00004-4

4.1 Background

As it was concluded in the previous chapter, probably the most viable approach to predict the structure's reliability or its service life under future performance conditions is through probability-based techniques involving time-dependent reliability analyses.

By using these techniques a quantitative measure of structural reliability is provided to integrate information on design requirements, material and structural degradation, damage accumulation, environmental factors, and nondestructive evaluation technology. The technique can also investigate the role of in-service inspection and maintenance strategies in enhancing reliability and extending service life. Several nondestructive test methods that detect the presence of a defect in a pipeline tend to be qualitative in nature in which they indicate the presence of a defect but may not provide quantitative data about the defect's size, precise location, and other characteristics that would be needed to determine its impact on structural performance. None of these methods can detect a given defect with certainty. The imperfect nature of these methods can be described in statistical terms. This randomness affects the calculated reliability of a component.

Structural loads, engineering material properties, and strength-degradation mechanisms are random. The resistance, $R(t)$, of a structure and the applied loads, $S(t)$, both are stochastic functions of time. At any time, t, the safety limit state, $G(R, S, t)$, is (Melchers, 1999):

$$G(R, S, t) = R(t) - S(t) \tag{4.1}$$

Making the customary assumption that R and S are statistically independent random variables, the probability of failure resulting from Eq. (4.1), $P_f(t)$, is (Melchers, 1999):

$$P_f(t) = P[G(t) \leq 0] = \int_0^\infty F_R(x) f_s(x) dx \tag{4.2}$$

in which $F_R(x)$ and $f_S(x)$ are the probability distribution function of R and density function of S, respectively. Eq. (4.2) provides quantitative measures of structural reliability and performance, provided that P_f can be estimated and validated.

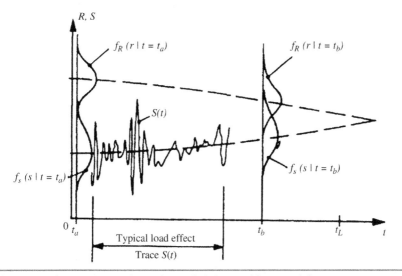

Figure 4.1 Schematic time-dependent reliability problem (Melchers 1999).

The probability that failure occurs for any one load application is the probability of limit state violation. Roughly, it may be represented by the amount of overlap of the probability density functions f_R and f_s in Fig. 4.1. Since this overlap may vary with time, P_f also may be a function of time.

As was mentioned in section 3.6 of Chapter 3, among the categories of reliability analysis methods, probabilistic methods should be considered for reliability analysis and service life prediction of pipes. For corrosion affected pipes, the method should also be time-dependent. The methods which have successfully been used previously include: first passageprobability method, gamma process concept method, and Monte Carlo simulationmethod. These techniques are practiced in Chapter 5 on case studies of different pipes.

4.2 First Passage Probability Method

The service life of a pipe or structure in general is a time period at the end of which the pipe stops performing the functions it is designed and built for. As was mentioned earlier, to determine the service life for pipes, a limit state function $(G(t) = R(t) - S(t))$ is introduced. Where $S(t)$ is the action (load) or its effect at time t and $R(t)$ is the acceptable limit (resistance) for the action or its effect. With the limit state function of Eq. (4.1), the probability of pipe (structural) failure, P_f, can be determined by:

$$P_f(t) = P[G(t) \leq 0] = P[S(t) \geq R(t)] \tag{4.3}$$

At a time that $P_f(t)$ is greater than a maximum acceptable risk in terms of the probability of pipe failure, P_a, it is the time the pipe becomes unsafe or unserviceable and requires replacement or repairs. This can be determined from the following:

$$P_f(T_L) \geq P_a \tag{4.4}$$

where T_L is the service life for the pipe for the given assessment criterion and acceptable risk. In principle, the acceptable risk, P_a, can be determined from a risk−cost optimization of the pipeline system during its whole service life. This can be further studied in Mann and Frey (2011) and Dawotola et al. (2012).

Eq. (4.3) represents a typical upcrossing problem in mathematics and can be dealt with using time-dependent reliability methods. Time-dependent reliability problems are those in which either all or some of the basic variables are modeled as stochastic processes. In this method, the structural failure depends on the time that is expected to elapse before the first occurrence of the action process $S(t)$ upcrossing an acceptable limit (the threshold) $L(t)$ sometime during the service life of the structure $[0, T_L]$. Equivalently, the probability of the first occurrence of such an excursion is the probability of failure $P_f(t)$ during that time period. This is known as "first passage probability" and can be determined by Melchers (1999):

$$P_f(t) = 1 - [1 - P_f(0)]e^{-\int_0^t v \, dt} \tag{4.5}$$

where $P_f(0)$ is the probability of structural failure at time $t = 0$ and v is the mean rate for the action process $S(t)$ to upcross the threshold $R(t)$.

The upcrossing rate in Eq. (4.5) can be determined from the Rice formula (Melchers 1999):

$$v = v_R^+ = \int_R^\infty (\dot{S} - \dot{R}) f_{S\dot{S}}(R, \dot{S}) d\dot{S} \tag{4.6}$$

where v_R^+ is the upcrossing rate of the action process S(t) relative to the threshold R, $\dot{R}$ is the slope of R with respect to time, $\dot{S}(t)$ is the time-derivative process of $S(t)$, and $f_{S\dot{S}}$ is the joint probability density function for S and $\dot{S}$. An analytical solution to Eq. (4.6) has been derived for a deterministic threshold R in Li and Melchers (1993) as follows:

$$v_R^+ = \frac{\sigma_{\dot{S}|S}}{\sigma_S} \varnothing\left(\frac{R-\mu_S}{\sigma_S}\right)\left\{\varnothing\left(-\frac{\dot{R}-\mu_{\dot{S}|S}}{\sigma_{\dot{S}|S}}\right) - \frac{\dot{R}-\mu_{\dot{S}|S}}{\sigma_{\dot{S}|S}}\Phi\left(-\frac{\dot{R}-\mu_{\dot{S}|S}}{\sigma_{\dot{S}|S}}\right)\right\} \quad (4.7)$$

where $\varnothing$ and Φ are standard normal density and distribution functions, respectively, μ and σ denote the mean and standard deviation of S and $\dot{S}$, represented by subscripts and "|" denotes the condition. For a given Gaussian stochastic process with mean function $\mu_S(t)$, and autocovariance function $C_{SS}(t_i, t_j)$, all terms in Eq. (4.7) can be determined, based on the theory of stochastic processes as detailed in Papoulis and Pillai (2002) as follows.

$$\mu_{\dot{S}|S} = E\left[\dot{S}\middle|S = R\right] = \mu_{\dot{S}} + \rho\frac{\sigma_{\dot{S}}}{\sigma_S}(R - \mu_S) \quad (4.8a)$$

$$\sigma_{\dot{S}|S} = [\sigma_{\dot{S}}^2(1-\rho^2)]^{1/2} \quad (4.8b)$$

where

$$\mu_{\dot{S}} = \frac{d\mu_S(t)}{dt} \quad (4.8c)$$

$$\sigma_{\dot{S}} = \left[\frac{\partial^2 C_{SS}(t_i, t_j)}{\partial t_i \partial t_j}\bigg|_{\cdot i=j}\right]^{1/2} \quad (4.8d)$$

$$\rho = \frac{C_{S\dot{S}}(t_i, t_j)}{\left[C_{SS}(t_i, t_i).C_{\dot{S}\dot{S}}(t_j, t_j)\right]^{1/2}} \quad (4.8e)$$

And the cross-covariance function is:

$$C_{S\dot{S}}(t_i, t_j) = \frac{\partial C_{SS}(t_i, t_j)}{\partial t_j} \quad (4.8f)$$

Because it is unlikely that the corrosion depth in a given pipe exceeds the wall thickness at the beginning of structural service, the probability of failure due to corrosion at $t = 0$ is zero, i.e., $P_f(0) = 0$. The solution to Eq. (4.5) can be expressed, after substituting Eq. (4.7) into Eq. (4.5), and considering that R is constant ($\dot{R} = 0$) therefore:

$$P_f(t) = \int_0^t \frac{\sigma_{\dot{S}|S}(t)}{\sigma_S(t)} \varnothing\left(\frac{R-\mu_S(t)}{\sigma_S(t)}\right)\left\{\varnothing\left(-\frac{\mu_{\dot{S}|S}(t)}{\sigma_{\dot{S}|S}(t)}\right) + \frac{\mu_{\dot{S}|S}(t)}{\sigma_{\dot{S}|S}(t)}\Phi\left(\frac{\mu_{\dot{S}|S}(t)}{\sigma_{\dot{S}|S}(t)}\right)\right\}d_\tau$$

$$(4.9)$$

The application of Eq. (4.9) for calculation of the failure probability for different case studies will be presented in sections 5.1.1 and 5.4.1.1 in Chapter 5.

■ 4.3 Gamma Process Concept

To deal with data scarcity and uncertainties, using stochastic models for time-dependent reliability analysis of deteriorating buried pipes can be considered. In order to model monotonic progression of a deterioration process, the stochastic gamma process concept can be used for modeling the reduction of pipe wall thickness due to corrosion. The gamma process is a stochastic process with independent, nonnegative increments having a gamma distribution with an identical scale parameter and a time-dependent shape parameter.

A stochastic process model, such as gamma process, incorporates the temporal uncertainty associated with the evolution of deterioration (e.g., Bogdanoff and Kozin, 1985; Nicolai et al., 2004; van Noortwijk and Frangopol, 2004).

The gamma process is suitable to model gradual damage monotonically accumulating over time, such as wear, fatigue, corrosion, crack growth, erosion, consumption, creep, swell, a degrading health index, etc. For the mathematical aspects of gamma processes, see Dufresne et al. (1991), Singpurwalla (1997), and van der Weide (1997).

Abdel-Hameed (1975) was the first to propose the gamma process as a model for deterioration occurring randomly in time. In his paper he called this stochastic process the "gamma wear process." An advantage of modeling deterioration processes through gamma processes is that the required mathematical calculations are relatively straightforward.

4.3.1 Problem Formulation

The mathematical definition of the gamma process is given in Eq. (4.10). Given that a random quantity d has a gamma distribution with shape parameter $\alpha > 0$ and scale parameter $\lambda > 0$ if its probability density function is given by:

$$Ga\left(d\middle|\alpha, \lambda\right) = \frac{\lambda^{\alpha}}{\Gamma(\alpha)} d^{\alpha-1} e^{-\lambda d} \tag{4.10}$$

Let $\alpha(t)$ be a nondecreasing, right continuous, real-valued function for $t \geq 0$, with $\alpha(0) \equiv 0$. $\Gamma(\alpha)$ denotes gamma function of α with mathematical definition of $\Gamma(\alpha) = (\alpha - 1)!$. The gamma process is a continuous-time stochastic process $\{d(t), t \geq 0\}$ with the following properties:

1. $d(0) = 0$ with probability one;
2. $d(\tau) - d(t) \sim Ga(\alpha(\tau) - \alpha(t), \lambda)$ for all $\tau > t \geq 0$;
3. $d(t)$ has independent increments.

Let $d(t)$ denote the deterioration at time t, $t \geq 0$, and let the probability density function of $d(t)$, in accordance with the definition of the gamma process, be given by

$$f_{d(t)}(d) = Ga\big(d\,|\,\alpha(t), \lambda\big) \qquad (4.11)$$

with mean and variance as follows:

$$E(d(t)) = \frac{\alpha(t)}{\lambda} \qquad (4.12)$$

$$Var(d(t)) = \frac{\alpha(t)}{\lambda^2} \qquad (4.13)$$

A pipe is said to fail when its corrosion depth, denoted by $d(t)$, is more than a specific threshold (a_0). Assuming that the threshold a_0 is deterministic and the time at which failure occurs is denoted by the lifetime T. Due to the gamma distributed deterioration, Eq. (4.11), the lifetime distribution can then be written as:

$$F(t) = \Pr(T \leq t) = \Pr(d(t) \geq a_0) = \int_0^{a_0} f_{d(t)}(d)d_d = \frac{\Gamma(\alpha(t), a_o\lambda)}{\Gamma(\alpha(t))} \qquad (4.14)$$

where $\Gamma(\nu, x) = \int_{t=x}^{\infty} t^{\nu-1}e^{-t}dt$ is the incomplete gamma function for $x \geq 0$ and $\nu > 0$.

To model corrosion in a pipe, in terms of a gamma process, the question that remains to be answered is how its expected deterioration increases over time. The expected corrosion depth at time t may be modeled empirically by a power law formulation (Ahammed and Melchers, 1997):

$$\alpha(t) = ct^b \qquad (4.15)$$

for some physical constants $c > 0$ and $b > 0$.

Because there is often engineering knowledge available about the shape of the expected deterioration in terms of the exponential parameter b in Eq. (4.15), this parameter may be assumed constant. The typical values for b from some examples of expected deterioration according to a power law are presented in Table 4.1.

The reliability analysis approach which is developed in this section by using the gamma process concept is entitled the "Gamma Distributed Degradation, GDD" model.

In the event of expected deterioration in terms of a power law (i.e., Eq. 4.15), the parameters c and λ can be estimated by using statistical estimation methods. The estimation procedure is discussed for the two scenarios including a case with available corrosion depth data and a case of unavailability of corrosion depth data.

TABLE 4.1 Typical Values for Exponential Parameter, b, in Different Deterioration Types

Deterioration type	Exponential parameter, b	References
Degradation of concrete due to reinforcement corrosion	1	Ellingwood & Mori (1993)
Sulfate attack	2	Ellingwood & Mori (1993)
Diffusion-controlled aging	0.5	Ellingwood & Mori (1993)
Creep	1/8	Cinlar et al. (1977)
Expected scour-hole depth	0.4	Hoffmans & Pilarczyk (1995) and van Noortwijk & Klatter (1999)

4.3.2 DEVELOPING GAMMA DISTRIBUTED DEGRADATION MODEL WITH AVAILABLE CORROSION DEPTH DATA

In this section using the gamma distributed degradation (GDD) model for reliability analysis of corrosion affected pipes in case of availability of corrosion depth data is discussed. The data of corrosion depth can be achieved by periodical inspections.

To model the corrosion as a gamma process with shape function $\alpha(t) = ct^b$ and scale parameter λ, the parameters c and λ should be estimated. For this purpose, statistical methods are suggested. The two most common methods that can be used for parameter estimation are the maximum likelihood and method of moments. Both methods for deriving the estimators of c and λ were initially presented by Cinlar et al. (1977) and were developed by van Noortwijk and Pandey (2003).

4.3.2.1 Maximum Likelihood Estimation

In statistics, maximum-likelihood estimation (MLE) is a method of estimating the parameters of a statistical model. When applied to a data set and given a statistical model, MLE provides estimates for the model's parameters.

In general, for a fixed set of data and underlying statistical model, the method of maximum likelihood selects values of the model parameters that produce a distribution that gives the observed data the greatest probability (i.e., parameters that maximize the likelihood function). Given that n observations are denoted by $x_1, x_2, \ldots, x_n$, the principle of maximum likelihood assumes that the sample data set is representative of the population. This has a probability density function of $f_x(x_1, x_2, \ldots, x_n; \theta)$, and chooses that value for θ (unknown parameter) that most likely caused the observed data to occur, i.e., once observations $x_1, x_2, \ldots, x_n$ are given, $f_x(x_1, x_2, \ldots, x_n; \theta)$ is a function of θ alone, and the value of θ that maximizes the above probability density function is the most likely value for θ.

In the current study a typical data set consists of inspection times $t_i, i = 1, \ldots, n$ where $0 = t_0 < t_1 < t_2 < \ldots < t_n$, and corresponding observations of the cumulative amounts of deterioration $d_i, i = 1, \ldots, n$ are assumed to be given as inputs of the model. Fig. 4.2 schematically shows a time-dependent degradation model in the case of two inspections with a deterministic path.

The maximum-likelihood estimators of c and λ can be determined by maximizing the logarithm of the likelihood function of the increments. The likelihood function of the observed deterioration increments $\delta_i = d_i - d_{i-1}$, $i = 1, \ldots, n$ is a product of independent gamma densities (van Noortwijk and Pandey, 2003):

$$l(\delta_1, \ldots, \delta_n | c, \lambda) = \prod_{i=1}^{n} f_{d(t_i) - d(t_{i-1})}(\delta_i) = \prod_{i=1}^{n} \frac{\lambda^{c[t_i^b - t_{i-1}^b]}}{\Gamma(c[t_i^b - t_{i-1}^b])} \delta_i^{c[t_i^b - t_{i-1}^b]} e^{-\lambda \delta_i} \quad (4.16)$$

To maximize the logarithm of the likelihood function, its derivatives are set to zero. It follows that the maximum likelihood estimator of λ is:

$$\hat{\lambda} = \frac{\hat{c} t_n^b}{d_n} \quad (4.17)$$

where $\hat{c}$ must be computed iteratively from the following equation:

$$\sum_{i=1}^{n} [t_i^b - t_{i-1}^b] \{ \psi(\hat{c}[t_i^b - t_{i-1}^b]) - log\delta_i \} = t_n^b log\left(\frac{\hat{c} t_n^b}{d_n}\right) \quad (4.18)$$

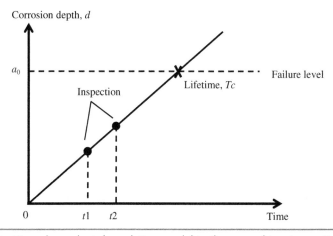

Figure 4.2 Time-dependent degradation model in the case of two inspections.

where the function $\psi(x)$ is the derivative of the logarithm of the gamma function:

$$\psi(x) = \frac{\acute{\Gamma}(x)}{\Gamma(x)} = \frac{\partial log\Gamma(x)}{\partial x} \quad (4.19)$$

4.3.2.2 Method of Moments

In statistics, the method of moments is a method of estimation of population parameters such as mean and variance by equating sample moments with unobservable population moments and then solving those equations for the quantities to be estimated. Assuming transformed times between inspections as $w_i = t_i^b - t_{i-1}^b, i = 1, \ldots, n$, the method-of-moments estimates c and λ from (van Noortwijk and Pandey, 2003):

$$\hat{c}\hat{\lambda} = \frac{\sum_{i=1}^{n} \delta_i}{\sum_{i=1}^{n} w_i} = \frac{d_n}{t_n^b} = \bar{\delta} \quad (4.20)$$

$$d_n \hat{\lambda} \left(1 - \frac{\sum_{i=1}^{n} w_i^2}{\left[\sum_{i=1}^{n} w_i \right]^2} \right) = \sum_{i=1}^{n} \left(\delta_i - \bar{\delta} w_i \right)^2 \quad (4.21)$$

The first equation from both methods (i.e., Eqs. 4.17 and 4.20) are the same and the second equation in the method of moments is simpler since it does not necessarily require iterations to find the unknown parameter ($\hat{c}$).

The flowchart in Fig. 4.3 illustrates the gamma distributed degradation model in the case of availability of corrosion measurements. To use this procedure, at least two measures of corrosion depth should be available for calculation of δ_i in Eqs. (4.18 and 4.21).

4.3.3 DEVELOPING GAMMA DISTRIBUTED DEGRADATION MODEL IN CASE OF UNAVAILABILITY OF CORROSION DEPTH DATA

In practice, most of the time for reliability analysis of corrosion affected pipes, data such as corrosion depth are not available. Therefore, a method should be developed for such cases of using the gamma distributed degradation model. As was mentioned in Section 4.4.1, in order to calculate the probability of failure over elapsed time (Eq. 4.14), the parameters corresponding to shape and scale parameters (α and λ) should be estimated. The steps for this purpose are:

a. Determining the approximate moments (mean and variance)
b. Estimating values for α and λ by using Eqs. (4.12 and 4.13)

Assuming $X_1, X_2, \ldots, X_n$ as basic random variables, moment approximation (i.e., step (a)) can be carried out by expanding the function $Y = Y(X_1, X_2, \ldots, X_n)$

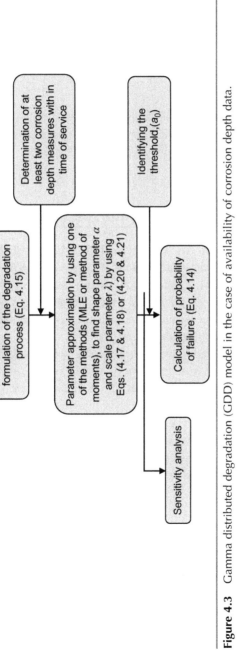

Figure 4.3 Gamma distributed degradation (GDD) model in the case of availability of corrosion depth data.

in a Taylor series about the point defined by the vector of the means $(\mu_{X_1}, \mu_{X_2}, \ldots, \mu_{X_n})$. By truncating the series, the mean and variance are (Papoulis and Pillai, 2002):

$$E(Y) \approx Y\left(\mu_{X_1}, \mu_{X_2}, \ldots, \mu_{X_n}\right) + \frac{1}{2}\sum_{i=1}^{n}\sum_{j=1}^{n}\frac{\partial^2 Y}{\partial X_i \partial X_j}\text{cov}(X_i, X_j) \tag{4.22}$$

$$\text{var}(Y) \approx \sum_{i}^{n}\sum_{j}^{n}c_i c_j \text{cov}(X_i, X_j) \tag{4.23}$$

The flowchart in Fig. 4.4 illustrates the gamma distributed degradation model in the case that corrosion measurements are not available. The procedure will be used for reliability analysis of case study pipelines in Chapter 5.

4.4 Monte Carlo Simulation Method

Monte Carlo simulation has been successfully used for reliability analysis of different structures and infrastructure (e.g., Camarinopoulos et al., 1999; Melchers, 1999; Sadiq et al., 2004; Yamini, 2009). Hence, the method can be used as a verification method to check the results which are obtained from the application of the two time-dependent analytical method (i.e., first passage probability method and gamma distributed degradation model).

Monte Carlo simulation techniques involve sampling at random to artificially simulate a large number of experiments and to observe the results. To use this method in structural reliability analysis, a value for each random variable is selected randomly ($\hat{x}_i$) and the limit state function ($G(\hat{x})$) is checked. If the limit state function is violated (i.e., $G(\hat{x}) \le 0$), the structure or the system has failed. The experiment is repeated many times, each time with randomly chosen variables. If N trials are conducted, the probability of failure then can be estimated by dividing the number of failures to the total number of iterations:

$$P_f \approx \frac{n(G(\hat{x}) \le 0)}{N} \tag{4.24}$$

The accuracy of the Monte Carlo simulation result depends on the sample size generated and, in the case when the probability of failure is estimated, on value of the probability (the smaller the probability of failure, the larger the sample size needed to ensure the same accuracy). The accuracy of the failure probability estimates can be checked by calculating their coefficient of variation (e.g., Melchers, 1999).

In order to improve the accuracy of estimating the probability of ultimate strength failure, while keeping the computation time within reasonable limits,

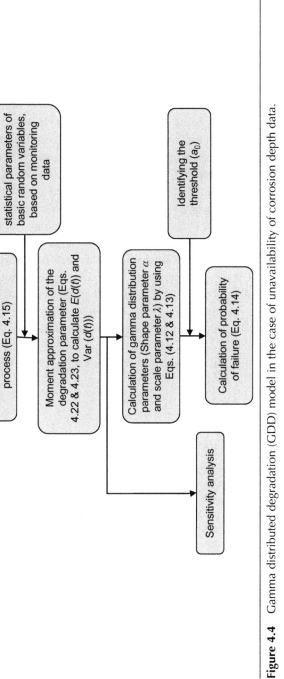

Figure 4.4 Gamma distributed degradation (GDD) model in the case of unavailability of corrosion depth data.

variance reduction techniques (e.g., importance sampling, Latin hypercube, and directional simulation) can be employed. However, in cases that the main emphasis is on serviceability failure, which can be estimated by a crude Monte Carlo simulation with very good accuracy within a relatively short computation time, such techniques are not necessary to be used (Val and Chernin, 2009).

Importance sampling is a variance reduction technique that can be used in the Monte Carlo method (Melchers, 1999). The idea behind importance sampling is that certain values of the input random variables in a simulation have more impact on the parameter being estimated than others. If these "important" values are emphasized by sampling more frequently, then the estimator variance can be reduced. Hence, the basic methodology in importance sampling is to choose a distribution which "encourages" the important values. The use of "biased" distributions will result in a biased estimator if it is applied directly in the simulation. However, the simulation outputs are weighted to correct use of the biased distribution, and this ensures that the new importance sampling estimator is unbiased.

The fundamental issue in implementing importance sampling simulation is the choice of the biased distribution which encourages the important regions of the input variables. Choosing or designing a good biased distribution is the "art" of importance sampling. The rewards for a good distribution can be significant runtime savings; the penalty for a bad distribution can be longer run times than for a general Monte Carlo simulation without importance sampling.

The details of the Monte Carlo method including sampling techniques can be found in Ditlevsen and Madesn (1996), Melchers (1999), and Rubinstein and Kroese (2008).

References

Abdel-Hameed, M., 1975. A gamma wear process. IEEE Trans. Rel. 24 (2), 152–153.

Ahammed, M., Melchers, R.E., 1997. Probabilistic analysis of underground pipelines subject to combined stress and corrosion. Eng. Struct. 19 (12), 988–994.

Bogdanoff, J.L., Kozin, F., 1985. Probabilistic Models of Cumulative Damage. John Wiley & Sons, New York.

Camarinopoulos, L., Chatzoulis, A., Frontistou-Yannas, S., Kallidromitis, V., 1999. Assessment of the time-dependent structural reliability of buried water mains. Rel. Eng. Sys. Safe. 65, 41–53.

Cinlar, E., Bazant, Z.P., Osman, E., 1977. Stochastic process for extrapolating concrete creep. J. Eng. Mech. Div.s 103 (EM6), 1069–1088.

Dawotola, A.W., Trafalis, T.B., Mustaffa, Z., van Gelder, P.H.A.J.M., Vrijling, J.K., 2012. Risk based maintenance of a cros-country petroleum pipeline system. J. Pipe. Sys. Eng. Prac Submitted September 2, 2011; accepted July 13, 2012.

Ditlevsen, O., Madsen, H.O., 1996. Structural Reliability Methods. John Wiley and Sons.

Dufresne, F., Gerber, H.U., Shiu, E.S.W., 1991. Risk theory with the gamma process. ASTIN Bull. 21 (2), 177–192.

Ellingwood, B.R., Mori, Y., 1993. Probabilistic methods for condition assessment and life prediction of concrete structures in nuclear power plants. Nuc. Eng. Des. 142, 155–166.

Hoffmans, G.J.C.M., Pilarczyk, K.W., 1995. Local scour downstream of hydraulic structures. J. Hyd. Eng. 121 (4), 326–340.

Li, C.Q., Melchers, R.E., 1993. Out-crossing from convex polyhedrons for non-stationary Gaussian processes. J. Eng. Mech. ASCE 119 (11), 2354–2361.

Mann, E., Frey, J., 2011. Optimized pipe renewal programs ensure cost-effective Asset management. Pipelines 2011, 44–54.

Melchers, R.E., 1999. Structural Reliability Analysis and Prediction, 2nd Edition John Wiley and Sons, Chichester.

Nicolai, R.P., Budai, G., Dekker, R., Vreijling, M., 2004. Modeling the deterioration of the coating on steel structures: a comparison of methods. Proceedings of the IEEE Conference on Systems, Man and Cybernetics. IEEE, Danvers, pp. 4177–4182.

Papoulis, A., Pillai, S.U., 2002. Probability, random variables, and stochastic processes, Fourth edition McGraw-Hill, New York, p. 852.

Rubinstein, R.Y., Kroese, D.P., 2008. Simulation and the Monte Carlo Method, second edition John Wiley and Sons.

Sadiq, R., Rajani, B., Kleiner, Y., 2004. Probabilistic risk analysis of corrosion associated failures in cast iron water mains. Reliab. Eng. Syst. Saf 86 (1), 1–10.

Singpurwalla, N., 1997. Gamma processes and their generalizations: an overview. In: Cooke, R., Mendel, M., Vrijling, H. (Eds.), Engineering Probabilistic Design and Maintenance for Flood Protection. Kluwer Academic Publishers, Dordrecht, pp. 67–75.

Val, D.V., Chernin, L., 2009. Serviceability reliability of reinforced concrete beams with corroded reinforcement. J. Struc. Eng. 135 (8), August 1.

van der Weide, H., 1997. Gamma processes. In: Cooke, R., Mendel, M., Vrijling, H. (Eds.), Engineering Probabilistic Design and Maintenance for Flood Protection. Kluwer Academic Publishers, Dordrecht, pp. 77–83.

van Noortwijk, J.M., Klatter, H.E., 1999. Optimal inspection decisions for the block mats of the Eastern-Scheldt barrier. Rel. Eng. Sys. Safe. 65 (3), 203–211.

van Noortwijk, J.M. and Pandey M.D., (2003), A stochastic deterioration process for time-dependent reliability analysis, Proceedings of the Eleventh IFIP WG 7.5 Working Conference on Reliability and Optimization of Structural Systems, 2-5 November 2003, Banff, Canada, pages 259-265.

van Noortwijk, J.M., Frangopol, D.M., 2004. Two probabilistic life-cycle maintenance models for deteriorating civil infrastructures. Prob. Eng. Mech. 19 (4), 345–359, p. 77–83.

Yamini, H., 2009. Probability of failure analysis and condition assessment of cast iron pipes due to internal and external corrosion in water distribution systems. University of British Colombia.

Further Reading

Institution of Civil Engineers, (2009), ICE State of The Nation Report Defending Critical Infrastructure, State of the Nation reports.

van Noortwijk, J.M., van der Weide, J.A.M., Kallen, M.J., Pandey, M.D., 2007. Gamma process and peaks-over-threshold distributions for time-dependent reliability. Rel. Eng. Sys. Safe. 92, 1651–1658.

Case Studies on the Application of Structural Reliability Analysis Methods

Abstract

The consequence of pipeline failure can be economically, socially, and environmentally devastating, which can cause massive costs for repair, enormous disruption of daily life, extensive pollution, and even human injuries. This necessitates a comprehensive procedure to estimate the likelihood of pipe failures and their remaining safe service life.

Strength of a pipe is affected by corrosion-induced reduction of the pipe wall. Therefore, it is of paramount importance to incorporate the effect of corrosion pit into the structural analysis of a pipeline. In this chapter reliability-based methodologies explained in Chapter 4 will be used for assessment of corroding pipes in four case studies. To cover a wide range of pipelines, case studies on oil and gas pipelines, steel water pipes, cast iron water pipes, and concrete sewers will be presented.

Chapter Outline

Reliability and Maintainability of In-Service Pipelines. DOI: https://doi.org/10.1016/B978-0-12-813578-5.00005-6

The use of theoretical methods for structural reliability analysis of in-service pipelines explained in Chapter 4 is practiced in this chapter through case studies. Different types of pipes including steel, cast iron, and concrete subject to different type of corrosive environment (i.e., oil and gas, water, and wastewater) are analyzed. The steps used for the reliability analysis in each case study include:

- Problem formulation: which introduces defining the limit state function(s), establishing the corrosion model and presenting the method for calculation of failure probability.
- Analysis of the results: which illustrates variation of failure probability versus time, verification of the results, and sensitivity analysis.

5.1 Case study 1-Oil and Gas Steel Pipes in USA

Oil and gas constitutes 60% of the world's fuel usage. Although pipelines are a very safe form of energy transportation, in the case of pipeline failures, the spilled oil and gas can cause a considerable hazard to the surrounding environment and population.

Deterioration and aging of pipeline infrastructure is one of the major problems facing the pipeline industry. More than half of the US oil and gas pipeline network is over 40 years old. 20% of Russia's oil and gas system is almost at the end of its design life and it is expected that in 15 years time 50% of their pipelines will be at the end of its design life. Metal corrosion is a common threat to the structural integrity of aging oil and gas pipelines. Corrosion as a time-dependent process gradually reduces the pipe strength and eventually causes the pipe failure. It has been shown (Sinha and Pandey, 2002; Anon, 2002; Thacker et al., 2010; Mahmoodian and Li, 2016a,b) that corrosion is the predominant cause for oil and gas pipe failures in many countries.

In Canada, for example, there are 34,000 km of oil pipelines and 26,000 km of gas pipelines where the prevention of corrosion-related failures at reasonable costs is the main concern (Sinha and Pandey, 2002). According to The World Factbook (2010), the US has approximately 800,000 km and Russia has 252,000 km of pipes transporting products like crude oil, natural gas, and petroleum products. The statistics for the UK and Australia are 20,000 km and 32,000 km, respectively. In the USA, corrosion has caused 23% and 39% of failures of oil and gas pipelines, respectively (Anon., 2002). One of the incidents caused by external corrosion was the leakage of an oil pipeline which spilled more than 140,000 gallons of crude on the Santa Barbara coast in May 2015.

In this section a case study on reliability analysis of oil and gas pipelines will be reviewed (Mahmoodian and Li, 2017). A stochastic model for pipe strength loss is developed which relates to key factors that affect the residual strength of a corroded pipe. A pipeline system fails when its residual strength falls below its operating pressure. The first passage probability method explained in Section 4.2 is employed to quantify the probability of failure due to corrosion so that the time for the pipeline to be failed, and hence require repairs, can be determined with confidence. To deal with the assessment of pipelines with more than one corrosion pit, system reliability analysis method is employed. Monte Carlo simulation technique is applied to verify the results of the analytical method. For an extensive reliability analysis, evaluation of the effect of various parameters on the pipeline structural reliability also is carried out via a parametric sensitivity analysis. The proposed methodology provides a rational and consistent approach to make quantitative assessment of pipeline failures.

A 2.5-kilometer length of crude oil pipeline in the US, made of grade X60 steel, is considered with 762 mm outer diameter (DN750) and 7.92 mm wall thickness. The maximum allowable operating pressure is 5.7 MPa. Inspection results have shown three pits caused by corrosion. The pipe characteristics, corrosion information, and the geometric information of the corrosion pits are presented in Tables 5.1 and 5.2.

5.1.1 PROBLEM FORMULATION

The practical failure criterion for corroded pipes is that residual strength falls below the operating pressure. As discussed in Section 3.2, this criterion can be expressed in the form of a failure function as follows:

$$G(Q, P_o, t) = Q(t) - P_o \tag{5.1}$$

where $Q(t)$ is the residual strength (structural resistance) at time t and P_o is the operating pressure (load effect). The residual strength $Q(t)$ decreases with time

TABLE 5.1 Values of Variables for Reliability Analysis in the Case Study

Symbol	Variable	Units	Mean	St dev.
D_o	Internal pipe diameter	mm	762	0
d	Pipe wall thickness	mm	7.92	0.077
P_o	Operating pressure	MPa	5.7	0
σ_y	Yield strength	MPa	461	16.13
c_d	Depth corrosion rate	mm/year	0.1	0
c_l	Length corrosion rate	mm/year	5.0	0

TABLE 5.2 Geometry of the Corrosion Pits in the Pipeline

	Pit 1		Pit 2		Pit 3	
	Length (mm)	Depth (mm)	Length (mm)	Depth (mm)	Length (mm)	Depth (mm)
Mean	95	2.2	120	1.9	165	1.5
Standard deviation	32	0.81	48	0.78	60	0.65

due to pipe deterioration, e.g., corrosion. With the failure function of Eq. (5.1), the failure probability can be determined from

$$P_f(t) = P[G(Q, P_o, t) \le 0] = P[Q(t) \le P_o] \tag{5.2}$$

where P indicates probability of an event.

The above equation represents a typical upcrossing problem, which can be dealt with using first passage probability method (Section 4.2). In a time-variant reliability problem some or all random variables are modeled as stochastic processes. For reliability problems involving the stochastic process of strength loss, as measured by residual strength $Q(t)$, the reliability depends on the expected elapsed time before the first occurrence of the stochastic process, $Q(t)$, upcrossing a threshold, P_o, sometime during the service life of the structure. Accordingly, the probability of the first occurrence of such an excursion is the failure probability, $P(t)$, during that period of time. This can be determined from Equation (4.5).

Following the procedure mentioned in Section 4.2, the failure probability can be determined from an Equation similar to Eq. (4.9). Replacing the corresponding parameters, the formulation for probability of failure will be:

$$P_f(t) = \int_0^t \frac{\sigma_{\dot{Q}|Q}(t)}{\sigma_Q(t)} \varnothing \left(\frac{P_o - \mu_Q(t)}{\sigma_Q(t)} \right) \left\{ \varnothing \left(-\frac{\mu_{\dot{Q}|Q}(t)}{\sigma_{\dot{Q}|Q}(t)} \right) + \frac{\mu_{\dot{Q}|Q}(t)}{\sigma_{\dot{Q}|Q}(t)} \Phi \left(\frac{\mu_{\dot{Q}|Q}(t)}{\sigma_{\dot{Q}|Q}(t)} \right) \right\} d_\tau$$

$$(5.3)$$

where $\dot{Q}$ is the time-derivative process of $Q(t)$; μ and σ are the mean and standard deviation of random variables represented by subscripts Q and $\dot{Q}$, and '|' denotes the condition. $\varnothing$ and Φ indicate standard normal density and distribution functions, respectively. According to the theory of stochastic processes (Melchers, 1999; Papoulis and Pillai, 2002), for a given Gaussian stochastic process with mean function $\mu_Q(t)$ and autocovariance function $C_{QQ}(t_i, t_j)$, the variables in the above equation can be determined as follows:

$$\mu_{\dot{Q}|Q} = E\left[\dot{Q}\Big|Q = P_o \right] = \mu_{\dot{Q}} + \rho \frac{\sigma_{\dot{Q}}}{\sigma_Q}(P_o - \mu_Q) \qquad (5.4a)$$

$$\sigma_{\dot{Q}|Q} = [\sigma_{\dot{Q}}^2(1 - \rho^2)]^{1/2} \qquad (5.4b)$$

where

$$\mu_{\dot{Q}} = \frac{d\mu_Q(t)}{dt} \qquad (5.4c)$$

$$\sigma_{\dot{Q}} = \left[\frac{\partial^2 C_{QQ}(t_i, t_j)}{\partial t_i \partial t_j} \Big|_{\cdot i=j} \right]^{1/2} \qquad (5.4d)$$

$$\rho = \frac{C_{Q\dot{Q}}(t_i, t_j)}{\left[C_{QQ}(t_i, t_i).C_{\dot{Q}\dot{Q}}\left(t_j, t_j\right) \right]^{1/2}} \qquad (5.4e)$$

and the cross-covariance function is:

$$C_{Q\dot{Q}}\left(t_i, t_j\right) = \frac{\partial C_{QQ}(t_i, t_j)}{\partial t_j} \qquad (5.4f)$$

In case of a pipe with more than one corrosion pit, the failure probability of the pipe can be estimated by using the systems reliability methods (Section 3.4). For the corroded pipe in this case study, the occurrence of failure for each corrosion pit will constitute its total failure. Therefore, a series system is more suitable for the failures assessment of corroded pipes. According to the theory of systems reliability the failure probability for a series system at time t, $(P_{f,s}(t))$, can be estimated by an equation the same as Eq. (3.7):

$$max\left[P_{f,i}(t)\right] \leq P_{f,s}(t) \leq 1 - \prod_{i=1}^{n} \left[1 - P_{f,i}(t)\right] \tag{5.5}$$

where $P_{f,i}(t)$ is the failure probability of the pipe due to the failure of ith corrosion pit on the pipe wall (determined by Eq. (5.3)) at time t and n is the number of corrosion pits existing in the pipeline.

At a time that $P_{f,s}(t)$ is greater than a maximum allowable risk in terms of the failure probability, P_a; (assuming the same consequences) it is the time the pipeline system is broken. This can be presented as follows:

$$P_{f,s}(T_f) \geq P_a \tag{5.6}$$

where T_f denotes the time the pipeline fails due to corrosion induced strength loss. In principle, P_a can be estimated from a risk−cost optimization analysis of the pipeline during its whole service life.

For Eq. (5.3) to be of practical use, i.e., determining the failure probability due to strength loss over time, the key is to develop a stochastic model for the residual strength. This is dealt with in the next section.

5.1.1.1 Model for Residual Strength and Corrosion

To determine the remaining strength of a corroded pipe, equations are mainly based on the ratio of the cross-section of the corrosion pit to the original cross-section. An analytical model developed by Kiefner and Vieth (1990), calculates the residual strength of a corroded pipe through the following equation:

$$Q = \frac{2d\sigma_f}{D_o}\left[\frac{1 - A/A_o}{1 - A/(MA_o)}\right] \tag{5.7}$$

where d is pipe wall thickness, D_o is pipe diameter, σ_f is the flow stress, A is the cross-section area of the corrosion pit projected onto the longitudinal axis of pipe, A_o is the original cross-section area before corrosion, and M is the Folias factor that accounts for bulging of the pipe before failure. The corroded cross-section area, A can be approximated as $A = a.l$, where a and l denote the average corrosion depth and longitudinal length of the corrosion pit, respectively.

The cross-section area before corrosion is $A_o = d.l$. Flow stress, σ_f, is generally defined as a function of the yield stress σ_y. Assuming $\sigma_f = 1.15\sigma_y$ and substituting for A and A_o in Eq. (5.7) results in the following equation (Nessim and Pandey, 1996; Cosham and Hopkins, 2004):

$$Q = \frac{2.3\sigma_y d}{D_o} \left[\frac{1 - a/d}{1 - a/(Md)} \right] \tag{5.8}$$

Nessim and Pandey (1996) define the Folias factor, M, as:

$$M = \sqrt{1 + 0.6275 \frac{l^2}{D_o d} - 0.003375 \frac{l^4}{D_o^2 d^2}} \text{ for } l^2/(D_o d) \leq 50 \tag{5.9a}$$

$$M = 0.032 \frac{l^2}{D_o d} + 3.3 \text{ for } l^2/(D_o d) > 50 \tag{5.9b}$$

As the corrosion pit grows with time, the residual strength given by Eq. (5.8) continues to decline. To predict the pipe strength at time t, corrosion growth rate needs to be estimated. The dimensions of corrosion pit at time t can be estimated by using a linear model for corrosion growth in steel pipes proposed by Sheikh and Hansen (1996):

$$a(t) = a(0) + c_d.t \tag{5.10}$$

and

$$l(t) = l(0) + c_l.t \tag{5.11}$$

where c_d and c_l denote the corrosion rate for pit depth and length, respectively. Considering the above equations, the residual strength of a corroded pipe would be

$$Q(t) = \frac{2.3\sigma_y d}{D_o} \left[\frac{1 - a(t)/d}{1 - a(t)/(M(t)d)} \right] \tag{5.12}$$

5.1.1.2 Stochastic Model for Residual Strength

The residual strength of a corroded pipe is a very random phenomenon. Therefore, it is justifiable to model the residual strength as a stochastic process, defined in terms of basic random variables as the primary contributing factors. Hence, the residual strength (i.e., Eq. (5.12)) can be presented as a function of basic random variables as well as time and can be expressed as:

$$Q(t) = f(l_o, a_o, \sigma_y, c_d, c_l, d, D_o, t) \tag{5.13}$$

where $l_o, a_o,$ and σ_y are the basic random variables. Assuming that the probabilistic information of the basic random variables are available, the statistical data of $Q(t)$ can be obtained by using numerical techniques such as Monte Carlo simulation.

The randomness of the residual strength can be taken to account by introducing a random variable, ξ_Q. This variable is defined in such a way that its mean is unity, i.e., $E(\xi_Q) = 1$ and its coefficient of variation, λ_Q, is a constant. Thus, Eq. (5.13) can be expressed as:

$$Q(t) = Q_c(t).\xi_Q \tag{5.14}$$

where $Q_c(t)$ is treated as a pure time function determined by residual strength equation (e.g., Eq. (5.12)). The mean and autocovariance functions of $Q(t)$ are (see, e.g., Li and Melchers, 2005)

$$\mu_Q(t) = E[Q(t)] = Q_c(t) \cdot E[\xi_Q] = Q_c(t) \tag{5.15}$$

$$C_{QQ}(t_i, t_j) = \lambda_Q^2 \rho_Q Q_c(t_i) Q_c(t_j) \tag{5.16}$$

where ρ_Q is autocorrelation coefficient for $Q(t)$ between two points in time t_i and t_j. With $\mu_Q(t)$ and $C_{QQ}(t_i, t_j)$, Eqs. (5.4a−5.4f) can be used to calculate other statistical parameters of $Q(t)$.

5.1.2 Results and Analysis

The probability of failure due to corrosion can be computed using the upper bound of Eq. (5.5) and the results are shown in Fig. 5.1 for different coefficients of correlation. Fig. 5.1 indicates that the effect of the autocorrelation of the fracture process between two points in time (i.e., ρ) on failure can be negligible. This may be of practical significance since ρ is not readily available and therefore

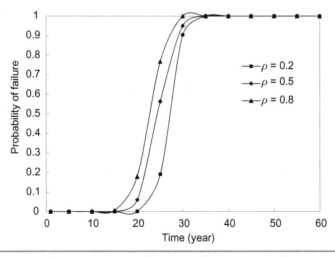

Figure 5.1 Probability of failure for different coefficients of correlation (ρ).

assumption of no correlation may not lead to significant difference. On the other hand, the theory of stochastic processes (Papoulis and Pillai, 2002) and the research experience (Li and Mahmoodian, 2013) suggest that the assumption of no autocorrelation between different time points generally leads to greater estimates of the probability of the occurrence of events, which is conservative for the assessment of pipeline deterioration.

The probability of failure due to corrosion with different acceptable operating pressure is shown in Fig. 5.2. There is a remarkable change in the safe life of the pipeline as the operating pressure increases. For example, for the acceptable failure probability of 0.05, the safe life increases from 15 to 25 years as the operating pressure decreases from 6.55 MPa to 4.85 MPa. This result can be used as a quantitative indicator by pipeline maintenance engineers to outline how and to what extent the operating pressure affects the safe life of the pipeline asset.

To verify the results of the first passage probability method in this case study, Monte Carlo simulation method was also used to calculate probability of failure. The details of the Monte carlo simulation method were discussed in Section 4.4. This comparison is shown in Fig. 5.3. As can be seen, the probability of failure as predicted by first passage probability method is in good agreement with that determined by Monte Carlo simulation, in particular in the region of small probabilities. As may be appreciated, it is the region of small probabilities, i.e., lower risk, that is of most practical interest.

The difference between the results of analytical and MC method can be due to approximations in the analytical method (specially determination of correlation coefficient ρ_Q and coefficient of variation λ_Q).

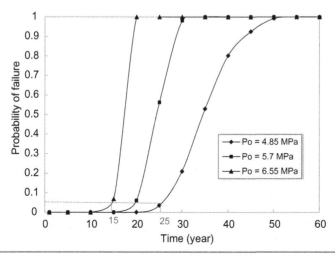

Figure 5.2 Probability of failure for different operating pressure.

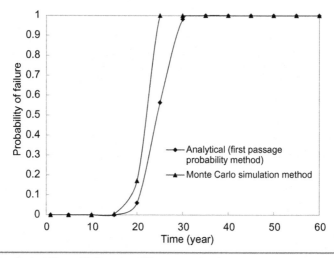

Figure 5.3   Probability of failure by different methods.

Since the costs of repairs are usually high for pipelines, it is of practical importance to accurately predict the time for repairs for corroded pipes. Therefore an optimum cost−benefit analysis can be applied for asset management of the pipeline system. In the current case study, the time for the pipe to fail, i.e., T_f, due to fracture of corrosion pits, can be determined for a given acceptable risk P_a. For instance, using results presented in Fig. 5.3 for the analytical method, it can be obtained that $T_f = 19$ years for $P_a = 0.05$. If there is no intervention during the service period of (0, 19) years for the pipe, such as maintenance and repairs, T_f represents the time for the first intervention or the end of service for the pipe for the given assessment criterion. The information of T_f (i.e., time for interventions) is of significant practical importance to pipeline engineers and asset managers of the oil and gas industry. It can guide the decision makers in prioritizing repairs and/or replacement of the pipes and also will help them to follow optimum maintenance strategies.

5.1.2.1 Sensitivity Analysis

In reliability analysis of a pipeline system, it is of interest to indicate those variables that affect the pipe failure most so that more effort can focus on those variables. For this purpose, a parametric study can be carried out on three variables that are deemed to be of significance to corrosion-induced pipe failures. These are (i) corrosion rate for pit depth and length as represented by c_d and c_l; (ii) geometry of the pipe as represented by the diameter and wall thickness, D_o and d; and (iii) pipe property as represented by yield strength, σ_y. The results are shown in Figs. 5.4−5.6.

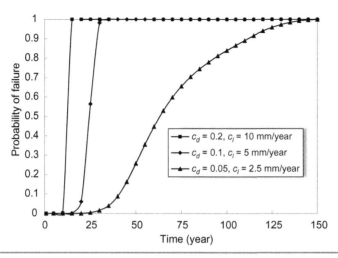

Figure 5.4 Effect of corrosion rates (c_d and c_l) on probability of failure.

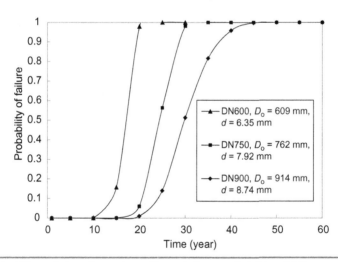

Figure 5.5 Effect of pipe dimensions (D_o and d) on probability of failure.

As can be seen from Fig. 5.4, corrosion rate is one of the important factors that affect the failure caused by corrosion. Doubling the rates of corrosion, e.g., $c_d = 0.2$ and $c_l = 10$, can lead to the reduction of failure time from 19 to 11 (years) given the same acceptance criterion. As may be appreciated, c_d and c_l can only be obtained from site-specific measurement on the pipe and its surrounding environment to be assessed. On the other hand, having lower rates of corrosion, e.g., $c_d = 0.05$ and $c_l = 2.5$, significantly increases the service life from 19 to 38 years. Therefore, accurate measurement of c_d and c_l is essential to predicting the reliability of corrosion affected steel pipelines.

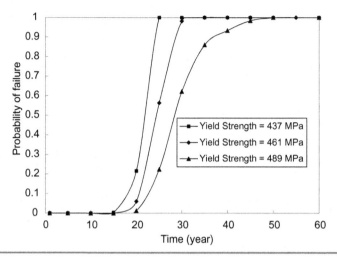

Figure 5.6 Effect of pipe yield strength (σ_y) on probability of failure.

Fig. 5.5 shows that the effect of the pipe geometry, as represented by the diameter and wall thickness, D_o and d, is moderate. As can be seen a two-step increase of the pipe size from DN600 to DN900 results in an increase of failure time from about 12 to 22 years under the same acceptance criterion. This is justifiable, as wall thickness is thicker for bigger pipes, therefore if all the other parameters are kept similar (e.g., internal pressure, corrosion rate, etc.), then there would be more sacrificial material for corrosion and consequently a longer time to lose wall thickness and structural integrity in bigger pipes. In comparison, the effect of yield strength on probability of failure can be negligible, especially in lower values of acceptable probability of failure which practically is the area of interest for safety assessment. This may be understandable since the failure is caused by the corrosion growth not the initial pipe strength.

It can be concluded that the proposed time-variant reliability method can be used as a rational tool for failure assessment of corrosion affected oil and gas pipelines with a view to determine the service life of the pipeline system.

5.2 Case Study 2-Mild Steel Water Pipes in Australia

As a case study, a steel distribution pipe in a water pipeline network built in 1975, located at East Sidney with 610 mm diameter and 9.53 mm thickness buried under a highway is selected. The pipe is circular and placed in a trench of 950 mm width and the depth of 3.10 m and the whole pipeline system is located above the groundwater level. The material of the pipe is steel with modulus of elasticity of 185 GPa. It is also assumed that the corrosion does not alter the dimensions of the internal pipe diameter.

5.2.1 PROBLEM FORMULATION

The start point of the structural reliability analysis is defining limit state functions (i.e., failure modes) as discussed in Section 3.3. A limit state is a condition of a structure beyond which it does not fulfill the relevant design criteria. Eq. (4.1) presented the general formulation for limit state function of a structure subject to a time varying process (e.g., corrosion). The failure occurs when the stress $(S(t))$ becomes greater than the resistance of the structure $(R(t))$.

When assessing the overall failure of a pipeline, it is important to consider the impact that each individual failure mode has on the pipeline system as a whole. The system failure analysis (see Section 3.4) is used to determine the interactions of the different limit states. Similarly, this type of failure analysis allows for the calculation of the overall system failure when the different failure modes are combined in parallel or series systems (Kołowrocki, 2008). For this case study, a series system failure analysis for estimating the probability of failure of steel water pipelines will be developed.

5.2.1.1 Limit State Functions

There are many known failure modes for buried pipelines. Identifying the dominating failure modes depends on the definition of physical model for the system, which involves consideration of loads or any other contributing parameters. Traditionally the most important parameters involved in the analysis of flexible buried pipes are loads, soil stiffness, and pipe stiffness (Moser and Folkman, 2008). External loadings and corrosion, which act through reduction of the pipe wall thickness, affect the failure condition of the pipeline.

The limit states to be considered in the failure analysis of steel water pipes have been presented in Table 5.3. These limit states are those which are controlled in the design stage of a buried flexible pipeline (Moser, 2010).

Nonuniform soil compaction along with overexcavation can be the reason behind nonuniform bedding, which leads to longitudinal deflection of the pipe. Flexible pipes are able to deform and move away from pressure. However, if the bending deflection exceeds the allowable longitudinal deflection threshold, the deflection failure will happen. Excessive bending can also result in flexural failure. It is also necessary to study the ring deflection of pipe and make sure it does not reach 5% of the inside diameter of pipe to prevent ring deflection failure (Moser and Folkman, 2008).

TABLE 5.3 Failure Modes and Determining Factors

LIMIT STATES	Formulation	Driving Factor
Flexural	$G(M_n, F_y, t) = M_n(t) - F_y$	Excessive bending due to ground movement and external loadings
Wall thrust	$G(T_a, T_{cr}, t) = T_a(t) - T_{cr}$	External loadings such as soil, traffic, and hydrostatic loads
Ring deflection	$G(\Delta X, D_i, t) = \Delta X(t) - 0.05 D_i$	External loads, mainly soil compression
Longitudinal deflection	$G(Y, Y_{max}, t) = Y(t) - Y_{max}$	Nonuniform bedding due to nonuniform material compaction along with over-excavation / corrosion
Pitting	$G(\Delta, W_t, t) = \Delta(t) - W_t$	Corrosion
Buckling	$G(P, P_{cr}, t) = P(t) - P_{cr}$	Elevated temperature / axial compression due to soil movement

Leakage usually happens when depth of pin-holes become greater than the thickness of the pipe wall. So it is important to consider the corrosion factors in choosing the wall thickness for a pipe. Buckling pressure and wall thrust are also considered as two vital factors which should be kept less than their critical thresholds to guarantee the safety of pipeline.

It should be noted that there are assumptions involved using some of formulations for limit state functions, for example, when calculating longitudinal deflection it is assumed that the pipe acts as a circular hollow section beam. Corrosion is also assumed to happen uniformly on the outer surface of the pipe. Moreover, this study considered that pipe segments survive or fail independently. Formularizations of the limit state criteria are detailed in Table 5.4.

5.2.1.2 Corrosion Model

Corrosion is a time-varying process and its rate changes over time. It starts off at a higher rate and then slows down with time. This is due to the protective properties of by-products, such as passive film, produced from the corrosion process (Mahmoodian and Alani, 2015). The following empirical formula represents the depth of corrosion in steel pipes (Li and Mahmoodian, 2013):

$$\Delta = at^b \tag{5.17}$$

For typical cases of corrosion experienced by pipelines, data is obtained from a regression analysis in existing literature and/or monitoring data, whereby the constants a and b were determined. In order to model the corrosion of steel in soil, a usually varies between 0.1 and 0.5 and value of b should be between 0.4 and 1.2 (Leygraf et al., 2016).

TABLE 5.4 Limit State Functions Formulation

	Failure Mechanism	Resistance	Reference	Stress	References
Ultimate Limit State	Flexural failure	$M_n = \frac{2D_t EA\gamma_0 S_{t1}}{D_m^2}$	Gabriel (2011)	F_y	Gabriel (2011)
	Wall thrust failure	$T_a = F_y(W_t - \Delta)\varphi$	Gabriel (2011)	$T_{cr} = 1.3(1.67P_s C_L + P_W)\frac{D_o}{2}$	Gabriel (2011)
Serviceability Limit State	Ring deflection failure	$\Delta X = \frac{K(D_L W_c + P_s)D_m}{\frac{8EI}{D_m^3}+0.061E'}$	BS 9295, 2010	$0.05D_i$	BS 9295, 2010
	Longitudinal deflection failure	$Y = \frac{P_s L^3}{250EI}$	Gorenc et al. (2005)	$Y_{max} = \frac{5L^3}{384EI}$	Gorenc et al. (2005)
	Leakage failure	$\Delta = aT^b$	Melchers (2009)	W_t	Melchers (2009)
	Buckling failure	$P = \frac{1}{S_f}\sqrt{\left(32R_w B' E_s \frac{EI}{D_m^3}\right)}$	Moser and Folkman (2008)	$P_{cr} = R_w \frac{W_c}{D_m} + \frac{P_s}{D_m}$	Moser and Folkman (2008)

TABLE 5.5 Values of Basic Variables, Worked Example

Symbol	Definition	Value	Mean	COV[a](%)	Distribution
E	Modulus of elasticity of pipe, GPa	---	185	1.5	Normal
E'	Modulus of soil reaction, kPa	---	1200	4.0	Normal
γ_s	Unit weight of soil, kN/m³	---	19.0	5.0	Normal
P_s	Wheel load, Live load, kPa	---	85.0	3.0	Normal
K	Bedding factor	---	11.0	1.0	Lognormal
W_t	Wall thickness, mm	---	9.53	1.0	Normal
r	Pipe radius, mm	---	309.3	1.0	Normal
H	Height of backfill, mm	---	3100.0	5.0	Normal
a	Multiplying constant	---	0.3	0.28	Normal
b	Exponential constant	---	0.65	0.32	Normal
F_y	Tensile strength, MPa	375	---	---	---
S_f	Safety factor for bending	1.5	---	---	---
S_f	Safety factor for buckling	2.5	---	---	---
D_L	Deflection lag factor	1.0	---	---	---
D_f	Shape factor	4.0	---	---	---
φ	Capacity factor for steel pipe	0.9	---	---	---
R_w	Water buoyancy factor	1.0	---	---	---
B_d	Maximum width of trench above pipe, m	2.5	---	---	---

[a]Note: COV stands for coefficient of variation.

It is important to note that because of the randomness associated with the variables in Table 5.5, the limit state parameters such as Marston's load on pipe, hydrostatic pressure and moment of inertia per unit length of pipe etc. will also have a degree of randomness.

5.2.1.3 Calculation of Failure Probability

In analyzing the reliability of the mentioned pipeline, a Monte Carlo Simulation is performed. First of all, the probability of failure in regard to time (t) is estimated for each of the limit state criteria. Then, by categorizing failure modes into two groups, namely serviceability and ultimate strength, and combining them in series or parallel systems, the overall failure probability of the system is calculated. Serviceability limit states are considered as a parallel system; because violation of them individually does not fail the whole system; while, ultimate limit

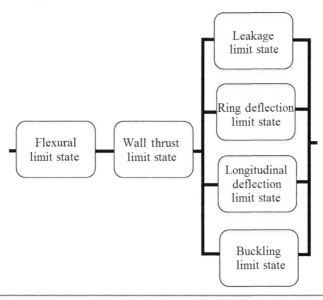

Figure 5.7   Multifailure system analyses.

states must be classified under series system criteria, since violation of each will cause the failure of the whole system (Fig. 5.7).

The following formulation represents the overall probability of failure of the system:

$$P_f(t) = 1 - \left[\left(1 - P_{f1}\right)\left(1 - P_{f2}\right)\left(1 - P_{f3}P_{f4}P_{f5}P_{f6}\right) \right] \tag{5.18}$$

where P_{f1} is the probability of flexural failure, P_{f2} is the probability of wall thrust failure, P_{f3} is the probability of ring deflection failure, P_{f4} is the probability of pitting failure, P_{f5} is the probability of buckling failure, and P_{f6} is the probability of longitudinal deflection failure.

For an inclusive reliability analysis, the effects of the stochastic variables on the system failure of the pipeline are analyzed by performing a sensitivity analysis of involved parameters.

Given that these variables are involved in the pipeline corrosion process and the limit state functions, finding the variables that affect the failure most is of utmost importance.

5.2.2 Results and Analysis

In this case study, the number of trials for Monte Carlo simulation is 10,000. The selected number of trials guarantees the smoothness of the estimated probability trend in Figs. 5.8–5.12 and eventually the accuracy of the estimations. A further

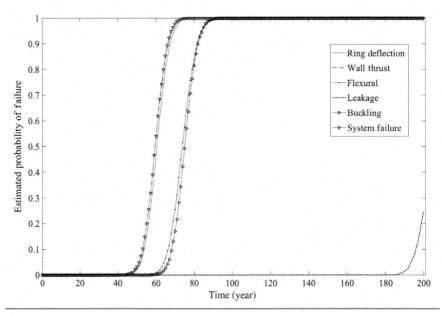

Figure 5.8 Probability of failure for individual and system limit state functions.

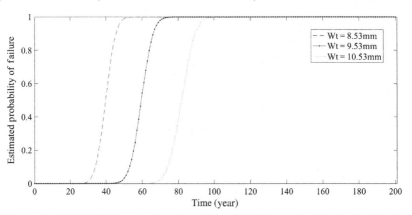

Figure 5.9 Sensitivity analysis for different wall thicknesses.

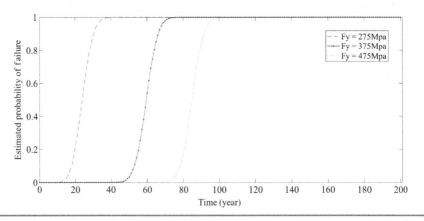

Figure 5.10 Sensitivity analysis for different yield strengths.

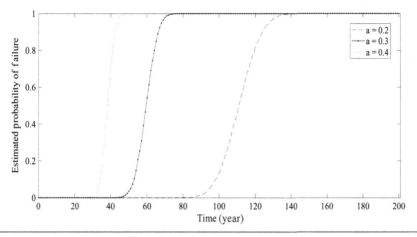

Figure 5.11 Sensitivity analysis for multiplying constant.

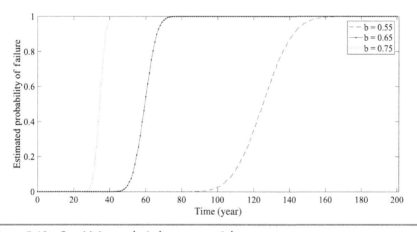

Figure 5.12 Sensitivity analysis for exponential constant.

increase in the number of trials does not necessarily improve the observable smoothness of the trend. However, more trials increase the computational burden of the simulation.

Fig. 5.8 shows the results of failure probability for each failure mode and also the system failure. From this figure the contribution of each limit state to the pipeline failure is apparent. The acceptable threshold for the estimated probability of failure can be assumed as 5%; it guarantees the safety of pipeline for 95% of its lifetime. Assuming a design life of 50 years for this pipeline, none of the limit states are violated before 50 years. Wall thrust and flexural are the first limit states that fail on the 60th and 61st year of the service life, respectively.

While evaluating the probability of failure for a pipeline, it is important that the various probabilities associated with the violation of the individual limit states are combined into a system analysis. In Fig. 5.8, the benefit of this approach of reliability analysis is apparent as the probability of pipeline system failure exceeds each of the individual limit state probabilities at every point. This is especially important during the initial 50 years of the graph, which is indicative of a typical pipeline's service life. It is evident from Fig. 5.8 that considering each failure mode individually in the reliability analysis of pipelines can mislead the reliability analysis, since it ignores the impact of the other limit states involved. For instance, considering only the leakage limit state, the service life of the pipeline is estimated more than 200 years, while it takes only 59 years for the system failure to happen (assuming acceptable failure probability of 0.05). Additionally, it can be seen that, the system failure is highly affected by wall thrust and flexural failure.

Using the results of Fig. 5.8, one can determine the time of failure (t_c) based on accepted probability of failure (P_a). Ascertainment of t_c is critical for design engineers and asset managers, as they must decide whether to repair or replace the structure in accordance with their budget allocation and risk analysis (Mahmoodian and Li, 2016a,b). For example, using the criteria of system failure, it can be obtained from Fig. 5.8 that, for an acceptable estimated probability of failure of 5%, the time of pipeline system failure is 59 years. If there is no repair or maintenance action during the service period of 59 years for the pipe, the pipe will not be serviceable after 59 years.

5.2.2.1 Sensitivity Analysis

The effect of the random variables on the failure of the pipeline can be analyzed using a parametric sensitivity analysis. As previously mentioned, these variables are involved in the corrosion process and thus, the parameters associated with determining the probability regarding the violation of the limit state functions. Consequently, determining the variables that influence this process the most, is of high importance. In order to study the effect of a variable using sensitivity analysis, first a deterministic value is given to the selected variable (i.e., its mean value) while the rest of the variables remain random. Then an upper and a lower value are given to the variable to observe the changes in failure probability curves.

The results of this parametric sensitivity analysis for four of the random variables are presented in Figs. 5.9–5.12. It should be noted that, these figures represent the estimated probability of failure of the pipeline system and not individual limit states. For instance, considering an acceptable probability of failure of 5%, it can be seen that having a wall thickness 1 mm less than that of the nominal value can reduce the lifetime of the pipeline system by more than 19 years (Fig. 5.9). This is expected to an extent, given that the thinner the pipe wall, the sooner depth of corrosion reaches the wall thickness of the pipe. Consequently, should there be any defects along the pipe wall an increased depth of corrosion is likely to lead to failure. Regardless of the limit state it is being designed for, the impact of wall thickness upon the system failure of the pipeline is significant (Fig. 5.9). Generally, choosing a wall thickness that is changed from its nominal value to either a higher or lower value drastically changes the service life of the pipeline.

The effect of mechanical properties of steel of pipe on pipe system failure is also of interest. Therefore, two more different values for the yield strength of the selected pipeline have been studied. Considering the same experimental condition and 5% acceptable probability of failure, a high strength structural steel water pipe with nominal yield strength of 475 MPa can operate safely up to 61 years longer than a steel pipe with 275 MPa tensile strength (Fig. 5.10). From this result, it can be seen that using a higher grade of steel for water pipes (i.e., higher yield strength) is generally beneficial; however, this would be more costly in practice.

In order to study the effect of corrosion on the service life of the pipeline, the sensitivity analysis must be applied to both the corrosion factors used as part of the nonlinear corrosion model in Eq. (5.18). As can be seen from the results illustrated in Figs. 5.11 and 5.12, the multiplication factor a has less impact on pipeline failure than its exponential counterpart b. For example, a decrease in a for an allowed probability of 0.05 increases the time of failure by 52 years in comparison with the nominal value (Fig. 5.11). For the same scenario, a 0.5 increase to the exponential factor b decreases the time to failure by approximately 66 years, proving that it is highly influential on pipeline failure (Fig. 5.12). It reveals that corrosion is a significant factor when considering the design of pipelines with long service lives.

Amongst the four random variables tested, the effect of a and b on the failure of pipelines is quite remarkable. The difference shown on each of these graphs for these variables indicates that the sensitivity of failure for studied pipelines is dependent on the values of these parameters. Accordingly, greater care should be taken when selecting the values for these variables in future studies.

Overall, this study indicates the importance of random variables on the system failure of buried steel water pipelines, as well as the considerable impact that the corrosion process can have over time. Similarly, existing literature also forms relationships between the probability of pipeline failure and time. Where this study differs is through applying a time-dependent deterioration method to the limit states set out in pipeline design manuals. The proposed methodology utilized the multifailure concept in order to fill the existing gap in similar researches, since most of them have considered only one failure mode when studying the reliability analysis of pipes. Moreover, the interaction of corrosion, as an environmental factor, and loading has been covered to introduce a novel and realistic failure assessment method. In addition, categorizing failure modes into serviceability and ultimate strength allows for more practical use of the probability of serviceability failure and/or the probability of ultimate failure.

On the other hand, applying the time-dependent corrosion formula to the parameters in each limit state can approximate the amount that each limit state contributes to the overall reliability of a pipeline. Hence, this study continues previous research in this field by adapting existing pipeline design codes to deal with the uncertainty associated with the time-dependent process of corrosion.

Further research can be done to address the situation where the pipeline is being protected by a coating. Coatings and other corrosion protection methods decrease the overall rate of corrosion and eventually affect the probability of failure of pipeline due to corrosion.

In this case study, a system reliability analysis for corrosion affected buried steel water pipelines has been developed. Using the violation of the limit state design criteria as the failure modes, the effects of external loads, pipe material, and corrosion etc. were considered. A nonlinear corrosion model was adopted in order to model the loss of wall thickness due to corrosion. Consequently, a time-dependent system reliability model considering all the involved limit states was developed, so that the probability of the pipeline system failure within time was estimated.

Using the aforementioned system failure analysis, the overall likelihood of pipeline failure was then determined by using a Monte Carlo simulation approach. The results highlighted that the studied pipeline is not safe and the appropriate repair or replacement time must be chosen based on the accepted probability of failure (P_a) and also time of failure (t_c). Time of repair and/or rehabilitation of steel water pipelines is of important benefit to pipeline engineers and assess managers.

From the results, it can be seen that the corrosion coefficients are the most influential factors among all the studied factors. Therefore, it is crucial for the designers to consider the impact of corrosion, as the most determining factor in the failure of pipelines, in the design stage to improve the service life of the pipelines. Additionally, the results of the proposed method, allows asset managers to plan an accurate maintenance strategy for the pipeline. Since maintenance strategies mainly involve finding a balance between probability of failure and the cost of reducing the risk, an optimal maintenance strategy can be obtained using the probability of failure estimation from the presented method and cost information which is determined by stakeholders.

▌ 5.3 Case Study 3-Cast Iron Water Mains in the UK

In this section, the applicability of the gamma process concept introduced in Section 4.3 for time-dependent reliability analysis is tested on a case study involving cast iron water pipes in the UK.

Both internal and external corrosion are considered and the results are compared and discussed. For a more realistic and reliable outcome both strength failure and fracture failure, in hoop and axial planes, are considered.

A cast iron pipe of 254 mm diameter and effective length of 6.5 m located in South East London is considered as the case to illustrate the proposed method. Other data for the cast iron pipe required for calculation is presented in Table 5.6 (Mahmoodian and Li 2018). In total 15 random variables are involved in the problem and their statistical data have been presented in Table 5.6.

5.3.1 PROBLEM FORMULATION

5.3.1.1 Definition of the Limit State Functions

Cast iron pipes can fail in many modes which in general can be summarized into two categories: loss of strength due to the reduction of wall thickness of the pipes, and loss of toughness due to the stress concentration at the tips of cracks or defects. Even in one category there can be many mechanisms that cause failure. The strength failure can be caused by hoop stress or axial stress in the pipes. A review of recent research literature (Sadiq et al., 2004; Moglia et al., 2008; Yamini, 2009; Clair and Sinha, 2012) suggests that current research on pipe failures focuses more on loss of strength than loss of toughness. As was mentioned in Section 3.3.7(b), the literature review also revealed that in most reliability analyses for buried pipes, multifailure modes are rarely considered although in practice this is the reality. Therefore the aim of this section is to consider multifailure modes in reliability analysis and service life prediction for cast iron pipes. Both loss of strength and toughness of the pipe are considered. A system reliability method is employed in calculating the probability of pipe failure over time, based on which the service life of the pipe can be estimated. Sensitivity analysis is also carried out to identify those factors that affect the pipe behavior most.

Buried pipes are not only subjected to mechanical actions (loads) but also environmental actions that cause the corrosion of pipes. Corrosion related defects would subsequently cause fracture of cast iron pipes. In the presence of corrosion pit, failure of a pipe can be attributed to two mechanisms: (i) the stresses in the pipe exceed the corresponding strength; or (ii) the stress intensity exceeds fracture

TABLE 5.6 Values of Basic Random Variables Used in the Case Study

Symbol	Variable		Units	Min	Mean	St. Dev.	Max	References
P	Internal Pressure		MPa	0.35	0.45	0.1	0.7	EPB 276 (2004)
D	Inner diameter		mm	240	254	14.28	260	BS78-2 (1965)
d	Wall thickness		mm	–	16	0.7	–	BS78-2 (1965)
K_m	Bending moment coefficient		–	–	0.235	0.04	–	Sadiq et al. (2004)
C_d	Calculation coefficient		–	–	1.32	0.20	–	Sadiq et al. (2004)
B_d	Width of ditch		mm	–	625	125	–	AWWA C600, (2005)
E_P	Modulus of elasticity of pipe		MPa	–	105,000	15,000	–	BS78-2 (1965)
K_d	Defection coefficient		–	–	0.108	0.0216	–	Sadiq et al. (2004)
I_c	Impact factor		–	–	1.5	0.375	–	Sadiq et al. (2004)
C_t	Surface load coefficient		–	–	0.12	0.024	–	Sadiq et al. (2004)
F	Wheel load		N	30,000	412,000	20,000	100,000	Sadiq et al. (2004)
A	Pipe effective length		mm	–	6500	200	–	AWWA C600 (2005)
γ	Soil Unit weight		N/mm^3	–	18.85×10^{-6}	18.85×10^{-7}	–	Sadiq et al. (2004)
k	Multiplying constant	Internal corrosion	–	–	0.92	0.18	–	Marshall (2001)
		External corrosion	–	–	2.54	0.5	–	Marshall (2001)
n	Exponential constant	Internal corrosion	–	–	0.4	0.08	–	Marshall (2001)
		External corrosion	–	–	0.32	0.06	–	Marshall (2001)

toughness of the pipe. Based on these two failure modes, two limit state functions can be established as follows.

5.3.1.2 Strength Limit State

Rajani et al. (2000) developed a formula for total stresses in a buried pipe including both hoop and axial stresses (see Fig. 5.13):

$$\sigma_h = \sigma_F + \sigma_S + \sigma_L + \sigma_V \tag{5.19}$$

where σ_h is the total hoop or circumferential stress in the pipe, σ_F, σ_S, σ_L, and σ_V are the hoop stresses due to internal fluid pressure, soil pressure, frost pressure, and traffic stresses respectively.

Similarly the total axial or longitudinal stress in the pipe can be expressed as:

$$\sigma_a = \sigma_{Te} + \sigma_P + (\sigma_S + \sigma_L + \sigma_V)\nu_p \tag{5.20}$$

where σ_a is the total axial stress in the pipe, σ_{T_e} is the stress related to temperature difference, σ_P is the axial stress due to internal fluid pressure, ν_p is Poisson's ratio of pipe material. Details of the equations and references used for determining the above stresses are presented in Table 5.7.

In practice, a pipe is usually under both axial and hoop stresses (σ_a and σ_h). If the yield strength of the pipe material is σ_y, the two limit state functions for strength can be established as follows

$$\text{Hoop stress limit state: } G_1\left(\sigma_y, \sigma_h, t\right) = \sigma_y - \sigma_h(t) \tag{5.21}$$

$$\text{Axial stress limit state: } G_2\left(\sigma_y, \sigma_a, t\right) = \sigma_y - \sigma_a(t) \tag{5.22}$$

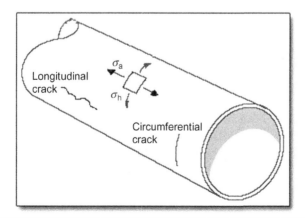

Figure 5.13 Stresses and cracks on a pipe wall (σ_h: hoop stress and σ_a: axial stress).

TABLE 5.7 Models for Stresses on Buried Pipes Considered in this Study

Stress Type	Model[a]	References
σ_F, hoop stress due to internal fluid pressure	$\frac{pD}{2d}$	Rajani et al. (2000)
σ_S, soil pressure	$\frac{3K_m\gamma B_d^2 C_d E_p d\ D}{E_p d^3 + 3K_d pD^3}$	Ahammed & Melchers (1994)
σ_V, Traffic stress	$\frac{3K_m I_c C_t F E_p d\ D}{A(E_p d^3 + 3K_d pD^3)}$	Ahammed & Melchers (1994)
σ_{T_e}, Thermal stress	$-E_p \alpha_p \Delta T_e$	Rajani et al. (2000)
σ_P, axial stress due to internal fluid pressure	$\frac{p}{2}\left(\frac{D}{d}-1\right)\nu_p$	Rajani et al. (2000)

[a]Notations are in Table 5.6.

5.3.1.3 Toughness Limit State

For localized stress concentration caused by defects, e.g., corrosion pits, the term stress intensity factor, K_I, is used (as it was mentioned in Section 2.2.2) to more accurately predict the stress state ("stress intensity") near the tip of a crack (caused by applied or residual stresses).

The formulations presented by Laham (1999) are used for calculation of stress intensity factors for crack pits in a pipe under different stresses. According to Laham (1999), the stress intensity factor for a crack pit in a pipe under hoop stress is as follows:

$$K_{I-h} = \sqrt{\pi a} \sum_{i=0}^{3} \sigma_i f_i\left(\frac{a}{d}, \frac{2c}{a}, \frac{R}{d}\right) \tag{5.23}$$

and the stress intensity factor for a crack pit in a pipe under axial stress:

$$K_{I-a} = \sqrt{\pi a}\left(\sum_{i=0}^{3} \sigma_i f_i\left(\frac{a}{d}, \frac{2c}{a}, \frac{R}{d}\right) + \sigma_{bg} f_{bg}\left(\frac{a}{d}, \frac{2c}{a}, \frac{R}{d}\right)\right) \tag{5.24}$$

where

K_{I-h} = Stress intensity factor for longitudinal crack in mode I, caused by hoop stress;

K_{I-a} = Stress intensity factor for circumferential crack in mode I, caused by axial stress;

a = Depth of the crack, i.e., corrosion pit;

σ_i = Stress normal to the crack plane;

f_i and f_{bg} = Geometry functions, depend on a, c (half-length of crack) and R (inner radius of pipe);

σ_{bg} = the global bending stress, i.e.; the maximum outer fiber bending stress

For internal and/or external crack pits, the difference in formulations of stress intensity factor (Eqs. 5.23 and 5.24) lies in geometry functions (i.e., f_i and f_{bg}), which have been presented in different tables by Laham (1999). Due to the propagation of corrosion, a changes with time so the stress intensity factors are time variant.

If K_C is the critical stress intensity factor, known as fracture toughness, beyond which the pipe cannot sustain propagation of the crack pit, the two limit state functions for fracture toughness can be established as follows:

$$\text{Axial fracture limit state} : G_3(K_C, K_{I-h}, t) = K_C - K_{I-h}(t) \qquad (5.25)$$

$$\text{Hoop fracture limit state} : G_4(K_C, K_{I-a}, t) = K_C - K_{I-a}(t) \qquad (5.26)$$

5.3.1.4 Corrosion Model

It has been well known that the predominant deterioration mechanism for cast iron pipes is electrochemical corrosion in the form of corrosion pits. Each spot of metal loss represents a corrosion pit that grows with time and reduces the thickness and mechanical resistance of the pipe wall. This process eventually leads to the collapse of the pipe.

As it was mentioned in Section 1.6.2, a number of models for corrosion of cast iron pipes have been proposed to estimate the depth of corrosion. There are debates in the research community as to whether the corrosion rate can be assumed linear or otherwise (e.g., Kucera and Mattsson, 1987; Sheikh and Hansen, 1996; Ahammed and Melchers, 1997; Rajani et al., 2000; Sadiq et al., 2004). The widely used model for corrosion pit is expressed in the form of the following:

$$a = kt^n \qquad (5.25)$$

where t is the exposure time and K and n are empirical coefficients which in practice are obtained by fitting the model to experimental data.

The modeling of corrosion pit is based on the experimental data from a UK Water Industry Research report (Marshall, 2001). This corrosion rate data used in this case study has been illustrated in Fig. 5.14. As the regression of available data fits a power law very well with high R-square value (Fig. 5.14, $R^2 = 0.959$ for internal corrosion and $R^2 = 0.857$ for external corrosion), the corrosion can be modeled, for both external and internal corrosion, as follows:

$$\mu_a = 2.54t^{0.32} \text{ for external corrosion} \qquad (5.26)$$

$$\mu_a = 0.92t^{0.4} \text{ for internal corrosion} \qquad (5.27)$$

where μ_a denotes the mean value for the depth of corrosion pit.

The widely used corrosion model (i.e., Eq. (5.25)) is selected for the multifailure mode reliability analysis of the cast iron pipe. Therefore the statistical values (mean and standard deviation) for k and n in Eq. (5.25) are again taken from the mathematical regression (Fig. 5.14) to the data from Marshall (2001). Based on

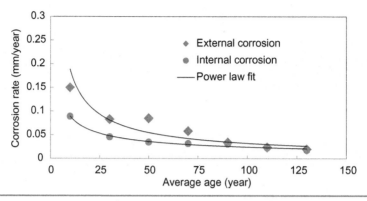

Figure 5.14 Corrosion rates for cast iron pipes. *(reproduced from Marshall (2001)).*

this data mean and standard deviation for corrosion coefficients (k and n) have been estimated (Table 5.6).

5.3.1.5 Calculation of Failure Probability

In the case of a pipe with multiple modes of failure, the probability of pipe failure can be determined using the methods of system reliability. Since the occurrence of either failure mode will constitute the failure of the pipe, a series system is appropriate for the assessment of pipe failures. The description of series, parallel, and complex systems have already been explained in Section 3.4.

Eq. (3.7) is used for calculation of the probability of series system failure. To estimate $P_{f_i}(t)$, the probability of failure due to the ith failure mode, GDD model is considered in this study.

Considering the methodology presented in Section 4.3, Eq. (4.14) is reproduced for the all four limit state functions (i.e., $G_1(\sigma_y, \sigma_h, t)$, $G_2(\sigma_y, \sigma_a, t)$, $G_3(K_C, K_{I-h}, t)$, and $G_4(K_C, K_{I-a}, t)$). The results are used as $P_{f_i}(t)$ for i = 1,...4 in Eq. (3.7) and consequently the probability of the pipe system failure ($P_{f,s}(t)$) is calculated.

5.3.2 RESULTS AND ANALYSIS

The results of the probability of the pipe system failure are presented in Figs. 5.15 and 5.16 for internal and external corrosion respectively.

The Monte Carlo simulation method (see Section 4.4 for the detail) is performed for verification of the results from GDD model. Figs. 5.17 and 5.18 show the comparison of the results of the probability of system failure from GDD model and Monte Carlo simulation for internal and external corrosion respectively.

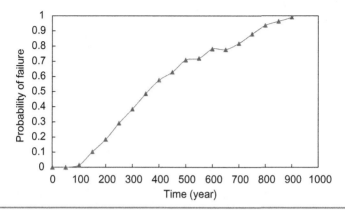

Figure 5.15   Probability of the pipe system failure (internal corrosion).

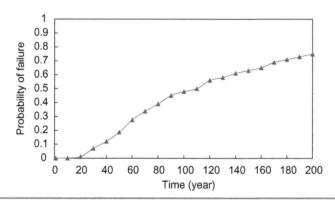

Figure 5.16   Probability of the pipe system failure (external corrosion).

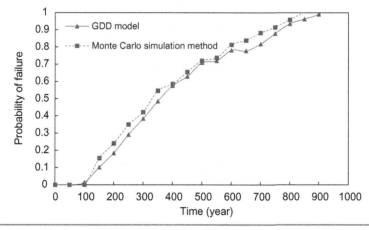

Figure 5.17   Verification of the results from GDD model by Monte Carlo simulation method (internal corrosion).

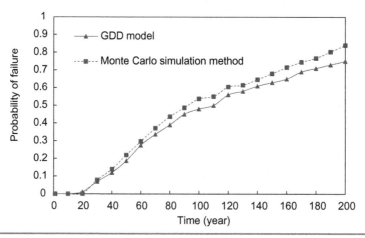

Figure 5.18 Verification of the results from GDD model by Monte Carlo simulation method (external corrosion).

The comparison shows that the probabilities of system failure predicted by GDD model can be verified by the results of Monte Carlo simulation method, particularly for small probabilities which are of most practical interest.

5.3.2.1 Sensitivity Analysis

It is known that the failure of a pipe can be affected by different factors, such as pipe geometry, corrosion coefficients, soil properties, and traffic loads. In view of the large number of factors that affect the corrosion process and failure modes, it is of practical significance to identify those factors that affect the failure most so that more research can focus on those factors. The effect of each variable on the pipe failure can be estimated by reliability based sensitivity analysis as it was outlined in Section 3.5.

To evaluate the sensitivity of the probability of failure to different random variables, sensitivity indexes are computed for the all 15 random variables. Figs. 5.19 and 5.20 show relative contribution (α^2) and sensitivity ratios (SR) for 25-year time steps, respectively.

It is obvious from the results that the sensitivity indexes of internal pressure (P), modulus of elasticity (E_P), deflection coefficient (K_d), impact factor (I_c), surface load coefficient (C_t), wheel load (F), and pipe effective length (A) are very low for all values of time.

Among all variables, the relative contributions and sensitivity ratios of the corrosion parameters (k and n) are highly remarkable. This indicates that corrosion is a very important factor for the design of underground pipelines with long lives.

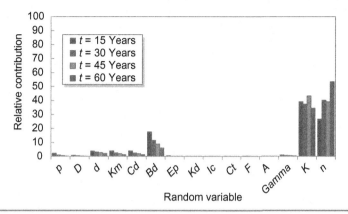

Figure 5.19 Relative contribution of random variables in pipe failure at different times for external corrosion.

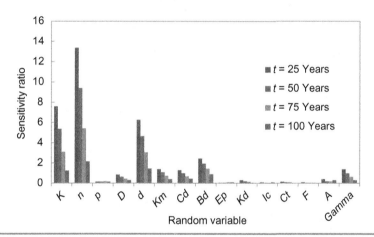

Figure 5.20 Sensitivity ratio of random variables subjected to external corrosion for different elapsed times.

Fig. 5.19 also shows that the relative contribution of some other variables (e.g., wall thickness (d), bending moment coefficient (K_m), calculation coefficient (C_d), and width of ditch (B_d)) is large at early ages, but it gradually decreases within time. This suggests the relative unimportance of these variables particularly for old pipes.

Further sensitivity studies were carried out to investigate the effect on probability of failure of the level of variability (i.e., coefficient of variation) of each of corrosion model coefficient (i.e., k and n) as major random variables. The coefficient of variation for each of these parameters was varied from 0 to 0.5 in steps of 0.1. The coefficient of variation of all other variables was kept constant at the values given in Table 5.6. Figs. 5.21 and 5.22 illustrate the results for external corrosion for four different points of time (t). Generally the probability of failure increases while the coefficient of variation increases.

A system reliability method for reliability analysis and service life prediction for a corrosion affected cast iron pipe in the UK was proposed. A merit of this method is that it considers multifailure modes as a system, including both loss of strength and toughness of the pipe in assessing pipe failures. For calculation of the probability of the pipeline failure, the GDD model was used and the results were verified by Monte Carlo simulation method. A sensitivity analysis was

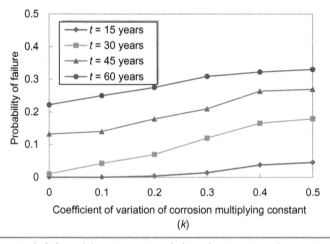

Figure 5.21 Probability of the pipe system failure due to external corrosion for different coefficients of variation of k at different times.

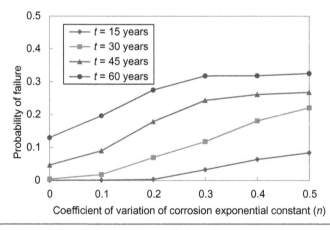

Figure 5.22 Probability of the pipe system failure due to external corrosion for different coefficients of variation of n at different times.

undertaken to identify the most significant factors among 15 random variables that affect the pipe behavior and failure. Among all variables, sensitivity indexes of the corrosion parameters (k and n) were highly remarkable. This indicates that corrosion is a very important factor for the design of underground pipelines with longer lives. High values of contribution for these two variables means that the sensitivity of these variables is more dependent on the actual value of their coefficient of variation. In such cases, more concern should be taken in order to determine relevant parameter values.

5.4 Case Study 4-Concrete Sewer Pipes in the UK

In this section the two model approaches outlined in Chapter 4 (i.e., first passage probability method and gamma distributed degradation model) are used for the reliability analysis and service life prediction of a concrete sewer pipeline in the UK. The results are verified by a Monte Carlo simulation and the weakness and strength of each method is discussed (Mahmoodian, 2013; Mahmoodian and Alani, 2013, 2014).

Two conditions are considered for the reliability analysis and service life prediction of the sewer pipeline in the UK. Firstly, the analysis is carried out based on an individual limit state function (i.e., considering one failure mode), and secondly, to have a more representative assessment of different failures to test reliability. The merit of this approach is the consideration of all possible corrosion induced failure modes that exist in practice. The effectiveness and contribution of the different variables (on the failure probability) provides new insight to the sensitivity analysis of concrete sewers.

In an individual failure analysis scenario, loss of concrete cover was assumed as the criterion for failure of the sewer. Four failure modes including flexural, shear, cracking, and cover loss were assumed as possible failure modes for multi-failure mode analysis. These modes of failure, which are set into the two categories of serviceability and ultimate strength limit states, were then put into a system configuration consisting of a combination of series and parallel systems.

5.4.1 INDIVIDUAL FAILURE MODE ASSESSMENT

The two developed methodologies in Chapter 4 for time-dependent reliability analysis of pipes are applied on CCTV data from surveys of concrete sewers in Harrogate in the United Kingdom. Harrogate is a spa town in North Yorkshire, England. The town is a tourist destination with a population of approximately 75,000. The study considers concrete sewers with a 500 mm diameter and 25 mm of internal concrete cover.

To incorporate the effect of increments in populations of residents and essentially the increase in the flow rate during the system's lifetime, the modeling assumes flow rates corresponding to relative depths (i.e., depth/diameter) of 0.2, 0.4, and 0.6, each occurring over a period of 25 years.

Reliability analysis and service life prediction of this concrete sewer pipeline is of interest for asset managers in North Yorkshire, England, to develop a risk-informed and cost-effective strategy in the management and maintenance of concrete sewers. The analysis can also help infrastructure managers to develop rehabilitation or replacement strategies for existing pipe networks with a view to better management of the pipe asset.

5.4.1.1 Problem Formulation

Limit State Function

According to the ASCE manual No.69 (1989), one of the performance criteria related to the stability of concrete sewers is to control the wall thickness reduction under an acceptable limit (normally concrete cover). In the theory of structural reliability this criterion can be expressed in the form of a limit state function as follows:

$$G(d_{max}, d, t) = d_{max}(t) - d(t) \qquad (5.28)$$

where:

d: Reduction in wall thickness due to corrosion (or corrosion depth), (mm)
d_{max}: Maximum permissible reduction in wall thickness (structural resistance or limit), (mm)
t: elapsed time

d_{max} may change with time although in most practical cases it has a constant value prescribed in design codes of practice and manuals.

Corrosion Model

The corrosion mechanism in concrete sewers was discussed in Section 1.6.1 and models for the rate of corrosion and wall thickness reduction were presented in the form of Equations 1.19 and 1.20, respectively.

Recalling Equation 1.20, the reduction in wall thickness in elapsed time t, is:

$$d(t) = c.t = 8.05k.(su)^{3/8}j.[DS] \times \frac{b}{P'A}.t \qquad (1.20)$$

where k is the factor representing the proportion of acid reaction, s is pipe slope, u is the velocity of stream (m/s), j is pH-dependent factor for proportion of H_2S,

[DS] is dissolved sulfide concentration (mg/L), A is acid-consuming capability or Alkalinity, b is the width of the stream surface, $\dot{P}$ is the perimeter of the exposed wall, and t is time.

To consider uncertainties about wall thickness reduction due to corrosion, a stochastic model is presented. Considering Equation 1.20, basic random variables affecting thickness reduction includes: $k, u, j, [DS], b/P'$, and A.

The wall thickness reduction due to corrosion is a function of basic random variables as well as time. It can be expressed as:

$$d(t) = f(k, u, j, [DS], b/P', A, t) \qquad (5.29)$$

where $k, u, j, [DS], b/P'$, and A are the basic random variables, the probabilistic information of which are presumed available.

Values for the basic random variables in the current case study are presented in Table 5.8.

Calculation of Failure Probability

With the limit state function introduced in the form of Eq. (5.28), the probability of failure of the concrete pipe due to the reduction of its wall thickness can be determined by:

$$P(t) = P[G(d_{max}, d, t) \leq 0] = P[d(t) \geq d_{max}(t)] \qquad (5.30)$$

The two developed methods for time dependent reliability analysis in Chapter 4 (i.e., first passage probability method and gamma distributed degradation model) are applied for calculation of the probability of failure of the concrete sewer case study in Harrogate in the UK.

TABLE 5.8 Statistical Data for the Basic Random Variables

Basic Variables	Units	Mean	Standard Deviation
K	–	0.8	0.1
j	–	0.2	0.04
[DS]	mg/L	1	0.5
u	m/s	0.6	0.1
b/P'	–	$h/D = 0.2$ ($b/P' = 0.36$)	0.11
		$h/D = 0.4$ ($b/P' = 0.55$)	0.18
		$h/D = 0.6$ ($b/P' = 0.71$)	0.23
A	–	0.2	0.06

Using First Passage Probability Method Eq. (5.30) is a typical upcrossing problem that can be solved by using first passage probability theory. In a time-dependent reliability problem all or some of the basic random variables are modeled as stochastic processes. For the above problem, the sewer failure depends on the time that is expected to elapse before the first occurrence of the stochastic process, $d(t)$, upcrosses a critical limit (the threshold, d_{max}) sometime during the service life of the sewer.

As it was described in Section 4.3, the probability of failure of a pipe can be determined by using first passage probability theory from Eq. (4.9). Considering Eq. (5.30) as the failure definition, reduction of wall thickness (d) is the action (load) and d_{max} is its effect or the acceptable limit (resistance). Therefore Eq. (4.9) can be reproduced with d replacing S and d_{max} replacing R.

$$P_f(t) = \int_0^t \frac{\sigma_{\dot{d}|d}(t)}{\sigma_d(t)} \varnothing \left(\frac{d_{max} - \mu_d(t)}{\sigma_d(t)} \right) \left\{ \varnothing \left(-\frac{\mu_{\dot{d}|d}(t)}{\sigma_{\dot{d}|d}(t)} \right) + \frac{\mu_{\dot{d}|d}(t)}{\sigma_{\dot{d}|d}(t)} \Phi \left(\frac{\mu_{\dot{d}|d}(t)}{\sigma_{\dot{d}|d}(t)} \right) \right\} d_\tau$$

(5.31)

For a given Gaussian stochastic process with mean function $\mu_d(t)$, and autocovariance function $C_{dd}(t_i, t_j)$, all terms in Eq. 5.31 can be determined as outlined in Section 4.3 by using the following formulations:

$$\mu_{\dot{d}|d} = E\left[\dot{d}|d = d_{max}\right] = \mu_{\dot{d}} + \rho_{\dot{d}} \frac{\sigma_{\dot{d}}}{\sigma_d}(d_{max} - \mu_d)$$ (5.32a)

$$\sigma_{\dot{d}|d} = [\sigma_{\dot{d}}^2(1 - \rho_{\dot{d}}^2)]^{1/2}$$ (5.32b)

where

$$\mu_{\dot{d}} = \frac{d\mu_d(t)}{dt}$$ (5.32c)

$$\sigma_{\dot{d}} = \left[\frac{\partial^2 C_{dd}(t_i, t_j)}{\partial t_i \partial t_j} \Big|_{i=j} \right]^{1/2}$$ (5.32d)

$$\rho_{\dot{d}} = \frac{C_{\dot{d}\dot{d}}(t_i, t_j)}{[C_{\dot{d}\dot{d}}(t_i, t_i) \cdot C_{\dot{d}\dot{d}}(t_j, t_j)]^{1/2}}$$ (5.32e)

and the cross-covariance function is

$$C_{\dot{d}\dot{d}}(t_i, t_j) = \frac{\partial C_{dd}(t_i, t_j)}{\partial t_j}$$ (5.32f)

$$C_{dd}(t_i, t_j) = \lambda_d^2 \rho_d \mu_d(t_i) \mu_d(t_j)$$ (5.32g)

where λ_d is the coefficient of variation of the wall thickness reduction which is determined based on Monte Carlo simulations and ρ_d is (auto-) correlation coefficient for the wall thickness reduction between two points in time t_i and t_j. Therefore all variables in Eq. (5.31) can be determined.

To estimate the probability of failure due to corrosion, a critical limit for the wall thickness reduction should be established. ASCE manual No.69 (1989) considers exposure of reinforcement as a criterion for failure. Therefore the maximum acceptable limit for wall thickness reduction (i.e., d_{max}) can be considered equal to the thickness of concrete cover ($d_{max} = a_0$, concrete cover).

Using Gamma Distributed Degradation (GDD) Model The average rate of wall thickness reduction within time in a concrete sewer is calculated through Equation 1.19 as explained in Section 1.6.1.2. Defining failure as the time when all concrete cover is corroded, the developed algorithm (GDD model) in Section 4.3.3 is used for formulation of the probability of failure based on the gamma process concept.

Given that a random deterioration (i.e., corrosion depth, d) has a gamma distribution with shape parameter $\alpha > 0$ and scale parameter $\lambda > 0$, formulation for calculation of the probability of failure for a corrosion affected pipe is developed as it was mentioned in Section 4.3.

The failure was defined as the time that all concrete cover on the reinforcement is corroded (ASCE manual No.69 (1989)), therefore the concrete sewer is said to fail when its corrosion depth, denoted by $d(t)$, is more than a special value (for instance, concrete cover, a_0). Assuming that the time at which failure occurs is denoted by the lifetime T, due to the gamma distributed deterioration, Eq. (4.14) can be used for calculation of the probability of failure.

$$F(t) = Pr(T \leq t) = Pr(\mathrm{d(t)} \geq a_0) = \int_0^{a_0} f_{d(t)}(d)d_d = \frac{\Gamma(\alpha(t), a_o\lambda)}{\Gamma(\alpha(t))} \qquad (4.14)$$

Where $\alpha(t) = ct^b$ is the shape parameter with physical constants $c > 0$ and $b > 0$ and λ is the scale parameter. The parameters $\alpha(t)$ and λ can be estimated by using the estimation method explained in Section 4.3.3. For the exponential parameter b, a value of one is assumed ($b = 1$) based on some examples of expected deterioration that have been presented in the Table 4.1.

5.4.1.2 Results and Analysis

The probability of failure due to wall thickness reduction is computed using first passage probability (Eq. 5.31) and the results are shown in Fig. 5.23. As can be concluded from this figure, the effect of the autocorrelation coefficient (ρ_d) on the probability of failure is negligible, specially for the area of interest (i.e., lower probability of failure).

Fig. 5.24 shows the results obtained for the probability of pipe failure by using the GDD method.

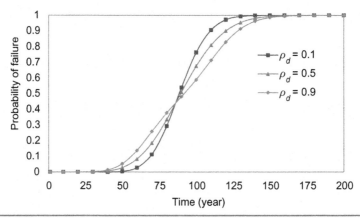

Figure 5.23 Probability of failure for different autocorrelation coefficient, ρ_d, from first passage probability method.

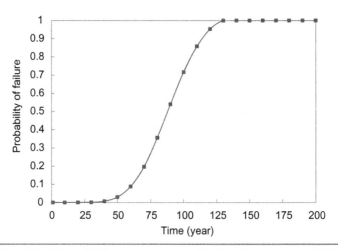

Figure 5.24 Probability of failure from gamma distributed degradation (GDD) model.

To verify the results obtained from the two developed methods in this section, Monte Carlo simulation method (see Section 4.4 for the detail) is used. The result of the probability of failure is presented in Fig. 5.25.

To be able to compare the results obtained from the two different methods, graphs in Figs. 5.23 and 5.24 are also illustrated in Fig. 5.25. The graph for first passage probability method is taken from the result for $\rho_d = 0.5$.

The comparison shows that the probabilities of failure predicted by the two methods are in good agreement and they can be verified by the results of the Monte Carlo simulation method, particularly for small probabilities which are of most practical interest.

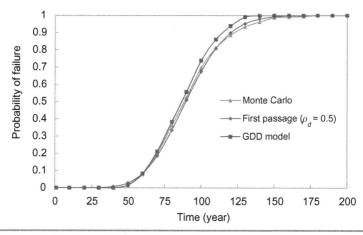

Figure 5.25 Verification of the results from the two methods by Monte Carlo simulation method.

Sensitivity Analysis

For a comprehensive reliability analysis it may be appropriate to assess the effect of different random variables on the service life of the sewer. It is of interest to identify those variables that affect the wall thickness reduction most, so that further studies can focus on those variables. For this purpose, relative contributions of variables to the variance of the limit state function were calculated using Eq. (3.8).

Fig. 5.26 illustrates sensitivities (α^2) or relative contribution of variables to the variance of the limit state function calculated based on the definition in Eq. (3.8) (Section 3.5). It can be concluded that while the contribution of some variables are relatively small, some other variables have considerable contribution to the limit state function. High values of contribution of variables mean that the sensitivity to those variables is more dependent on the actual value of their coefficient of variation. In such cases, more concern should be taken in order to determine relevant parameter values.

The GDD model was run to elaborate on the effect of variables with high values of relative contribution (i.e., [DS], A, and $b/\acute{P}$) on the probability of failure. As can be concluded from Fig. 5.27, a change in sulfide concentration has a considerable effect on service life of the concrete sewer.

Fig. 5.28 also shows the effect of sewage flow quantity on the service life of the sewer. The ratio of the width of the stream surface to the perimeter of the exposed wall ($b/\acute{P}$) has a significant effect on the corrosion rate of concrete and consequently the service life of the sewer decreases dramatically by increasing the ratio of $b/\acute{P}$.

To investigate more about the effect of concrete properties on the service life of concrete sewers, more studies on the effect of change in alkalinity (acid

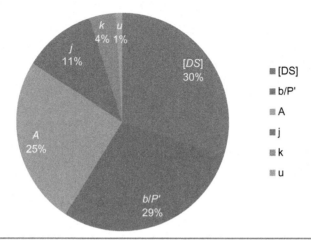

Figure 5.26 Relative contributions of random variables in failure function.

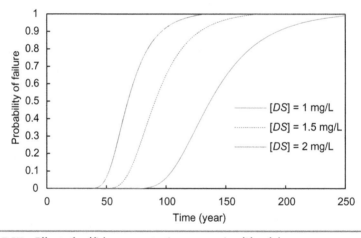

Figure 5.27 Effect of sulfide concentration on service life of the sewer.

consuming capability) of concrete on the probability of failure were carried out. The result of changing alkalinity from 0.14 to 0.22 is presented in Fig. 5.29.

Further sensitivity studies were carried out to investigate the effect on the reliability index of the level of variability (i.e., coefficient of variation) of each of the major random variables. The reliability index (β) was chosen for this work in preference to the probability of failure, mainly to facilitate the interpretation of the results. Although these two quantities are directly related (Eq. 3.5), the interpretation of results would be more appropriate when dealing with reliability index rather than probability of failure.

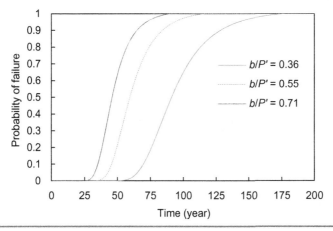

Figure 5.28 Effect of the ratio of the width of the stream surface to the perimeter of the exposed wall $(b/\dot{P})$ on service life of the sewer.

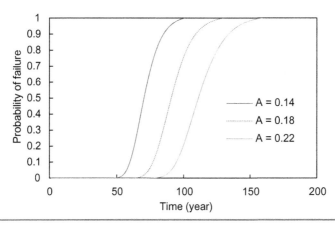

Figure 5.29 Probability of failure for different values of alkalinity, A.

The random variables chosen for this study were three significant variables in terms of the α^2 contribution in Fig. 5.28 (i.e., $[DS]$, A, b/P').

The coefficient of variation for each of these random variables was varied from 0 to 0.5 in steps of 0.1. The coefficient of variation of all other variables was kept constant at the values given in Table 5.8. Figs. 5.30−5.32 illustrate the results for four different pipeline lifespans (t). A period of service life from $t = 20$ year to $t = 35$ year which results in practical reliability indexes from 1.5 to 4.5 is selected for this study.

The results show that the reliability index decreases as time and the coefficient of variation of random variables increases. It is also observed that the variability

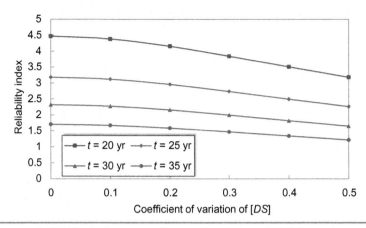

Figure 5.30 Reliability index vs. coefficient of variation of [*DS*] for various values of the pipeline elapsed time.

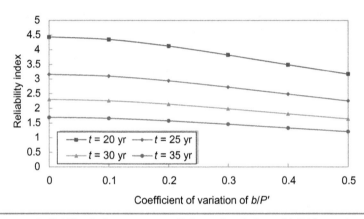

Figure 5.31 Reliability index versus coefficient of variation of *b*/*P*′ for various values of the pipeline elapsed time.

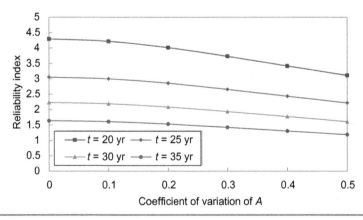

Figure 5.32 Reliability index versus coefficient of variation of alkalinity (*A*) for various values of the pipeline elapsed time.

of these three random variables for low values of t has a more significant effect on the reliability index.

5.4.2 MULTIFAILURE MODE ASSESSMENT

In this section, the case of the concrete sewer pipeline in Harrogate UK is analyzed by considering more than one possible failure mode. This study which is called multifailure mode reliability analysis will eventually end up with a more reliable service life estimation.

5.4.2.1 Problem Formulation

Limit State Functions
For concrete sewers, failure does not necessarily imply structural collapse (ultimate strength failure) but in most cases is indicated by loss of structural serviceability, as characterized by concrete cracking and/or concrete cover loss. In a comprehensive reliability analysis it is of interest to take into account both serviceability failure and ultimate strength failure.

In some cases of reliability analysis of a structure, various limit states such as bending, shear, cracking, and deflection may apply in a composition referred to as a "system." The detail of system reliability analysis was explained in Section 3.4. As was mentioned in that section, in a series system, attainment of any limit state constitutes failure of the structure. All components of a parallel system must fail for system failure to occur.

A concrete sewer can fail in multiple modes due to different limit state violations. Therefore, the probability of the sewer failure should be determined using the methods of systems reliability analysis.

Considering design codes of practice and manuals for reinforced concrete pipe design (ASCE 15−98, 2000; ASCE 60, 2007), the four following failure modes (limit state functions) should all be considered for buried concrete sewers:

$$\text{Flexural limit state: } G_1(M_u, M_s, t) = M_u(t) - M_s(t) \qquad (5.33a)$$

$$\text{Shear limit state: } G_2(V_b, V_s, t) = V_b(t) - V_s(t) \qquad (5.33b)$$

$$\text{Excessive crack limit state: } G_3(F, F_{cr}, t) = F(t) - F_{cr}(t) \qquad (5.33c)$$

$$\text{Cover loss limit state: } G_4(\Delta, a_o, t) = \Delta(t) - a_o(t) \qquad (5.33d)$$

where $M_u, V_b, F,$ and Δ are flexural strength, shear strength, crack control factor, and concrete thickness reduction, respectively, which are considered as thresholds for limit state functions. Formulization of these four resistance modes for a reinforced concrete pipe are presented in Table 5.9.

TABLE 5.9 Formulization of Different Resistance Modes of the Concrete Sewer

Failure Mechanism		Stationary Formulization	Source	Time Dependent Formulization
		Equation		
Strength	Flexu-ral Failure	$M_u = A_s f_y \left(d - \frac{a}{2}\right) + N_u \left(\frac{h-a}{2}\right)$	ASCE 15–98 (2000)	$M_u(t) = A_s f_y \left(d - \frac{a}{2}\right) + N_u \left(\frac{h(t) - a}{2}\right)$
	Shear Failure	$V_b = 0.083 b \varnothing_v d F_{vp} \sqrt{f'_c} \left(1.1 + 63 \times \frac{A_s}{bd}\right) \left[\frac{F_d F_N}{F_c}\right]$ $F_N = 1 + \frac{N_u}{3.5bh}$	ASCE 15–98 (2000)	$V_b(t) = 0.083 b \varnothing_v d F_{vp} \sqrt{f'_c} \left(1.1 + 63 \times \frac{A_s}{bd}\right) \left[\frac{F_d F_N(t)}{F_c}\right]$ $F_N(t) = 1 + \frac{N_u}{3.5bh(t)}$
Serviceability	Exces-sive Crack width	$F = \dfrac{B_1}{5250 \varnothing_f dA_s}$ $\left[\dfrac{M_s + N_s \left(d - \frac{h}{2}\right)}{ij} - \left(0.083 C_1 bh^2 \sqrt{f'_c}\right)\right]$ $j \cong 0.74 + 2.54 e/di = \dfrac{1}{1 - \frac{jd}{e}}$ $e = \dfrac{M_s}{N_s} + d - \dfrac{h}{2}$	ASCE 15–98 (2000)	$F(t) = \dfrac{B_1}{5250 \varnothing_f dA_s}$ $\left[\dfrac{M_s + N_s \left(d - \frac{h(t)}{2}\right)}{i(t)j(t)} - \left(0.083 C_1 bh(t)^2 \sqrt{f'_c}\right)\right]$ $j(t) \cong 0.74 + 2.54 \dfrac{e(t)}{d}$ $i(t) = \dfrac{1}{1 - \frac{i(t)d}{e(t)}}$ $e(t) = \dfrac{M_s}{N_s} + d - \dfrac{h(t)}{2}$
Loss of Concrete Cover		$\Delta = h - d - \frac{d_b}{2}$	ASCE 60 (2007)	$\Delta(t) = h(t) - d - \frac{d_b}{2}$

M_s, V_s, F_{cr} and a_o are flexural stress, shear stress, crack control limit and concrete cover The formulae presented by design codes for the resistance modes have stationary formats; while in the case of corrosion, the wall thickness of the pipe is a time-dependent parameter (i.e., decreases within the time). Hence, the time-dependent format of each formula is given in the last column of Table 5.9.

Symbols

a	depth of the equivalent rectangular stress block, (mm)
A	the acid-consuming capability of the wall material
A_s	Area of tension reinforcement in length b, (mm^2/m)
b	unit length of pipe, 1000 mm
B_1	crack control coefficient for effect of spacing and number of layers of reinforcement
c	the average rate of corrosion (mm/year)
C_1	crack control coefficient for type of reinforcement
d	distance from compression face to centroid of tension reinforcement, (mm)
d_b	diameter of rebar in inner cage, mm
$[DS]$	dissolved sulfide concentration (mg/L)
f'_c	design compressive strength of concrete, (MPa)
f_y	design yield strength of reinforcement, (MPa)
F	crack width control factor
F_c	factor for effect of curvature on diagonal tension (shear) strength in curved components
F_d	factor for crack depth effect resulting in increase in diagonal tension (shear) strength with decreasing d
F_N	coefficient for effect of thrust on shear strength
h	overall thickness of member (wall thickness), (mm)
i	coefficient for effect of axial force at service load stress
k	acid reaction factor
J	is pH-dependent factor for proportion of H$_2$S
w	the width of the stream surface
P'	perimeter of the exposed wall
M_s	service load bending moment acting on length b, (Nmm/m)
M_u	factored moment acting on length b, (Nmm/m)
N_s	axial thrust acting on length b, service load condition (+ when compressive, − when tensile), (N/m)
N_u	factored axial thrust acting on length b, (+ when compressive, − when tensile), (N/m)
s	is the slope of the pipeline
t	elapsed time
u	is the velocity of the stream (m/sec)
V_b	basic shear strength of length b at critical section
Φ	the average flux of H$_2$S to the wall
$\varnothing_f$	strength reduction factor for flexure
ϕ_v	strength reduction factor for shear
Δ	reduction in wall thickness due to corrosion, (mm)
Δ_{max}	maximum permissible reduction in wall thickness (structural resistance or limit), (mm)

The four limit states (i.e., Eqs. 5.33a−5.33d) can be classified in the two main categories of failure modes, namely serviceability limit states and ultimate

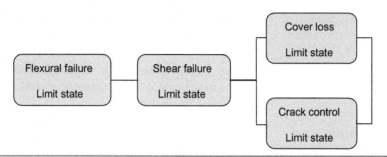

Figure 5.33 System combination of the four limit state functions for multifailure mode reliability analysis of the concrete sewer.

strength limit states. If a pipe loses its flexural strength and/or its shear strength it has completely failed. Therefore flexural limit state and shear limit state are considered as ultimate strength limit state functions. On the other hand, if a pipe cracks or loses its cover, it is not necessarily failed structurally, but it is failed from a serviceability point of view. Therefore, crack limit state and cover loss limit state can be considered as serviceability limit states.

As mentioned earlier, each failure mode happens when the limit state function is violated (i.e., $G_i \leq 0$). To consider all the four modes as a system, it is necessary to clarify the combination of the limit state functions. Fig. 5.33 presents the combination which is suggested for a system reliability analysis of the concrete sewer; it is a combination of series and parallel systems. The two serviceability limit states (crack and cover loss) are considered parallel, because violation of them individually does not fail the whole system. On the other hand violation of flexural limit state and/or shear limit state will cause the failure of the whole system and therefore these two limit states are set in a series combination.

5.4.3 CORROSION MODEL

The same corrosion model as presented in Eq. (1.20) with $k, u, j, [DS], b/P'$, and A as the basic random variables is considered for the multifailure mode reliability analysis of the concrete sewer.

5.4.4 CALCULATION OF FAILURE PROBABILITY

According to the theory of systems reliability, the probability of failure for a series system ($P_{fs}(t)$) can be estimated by Eq. (3.7):

$$\max[P_{f_i}(t)] \leq P_{fs(t)} \leq 1 - \prod_{i=1}^{m} \left[1 - P_{f_i}(t)\right] \tag{3.7}$$

where $P_{f_i}(t)$ is the probability of failure due to the ith failure mode of pipe and m is the number of failure modes considered in the system.

Considering the system configuration presented in Fig. 5.33 and the upper bound of Eq. (3.7), the probability of failure of the whole concrete sewer system can be calculated by the following equation:

$$P_{fs}(t) = 1 - \left(1 - P_{f_1}\right)\left(1 - P_{f_2}\right).\left(1 - P_{f_3}.P_{f_4}\right) \qquad (5.34)$$

where:

P_{f_1} is the probability of flexural failure;
P_{f_2} is the probability of shear failure;
P_{f_3} is the probability of crack failure;
P_{f_4} is the probability of cover loss failure.

For reliability analysis of the sewer, initially, the probability of failure in time t for each failure mode ($P_{f_i}(t)$, i = 1,..,4) is estimated by using one of the two proposed methods (first passage probability and/or GDD model). Then Eq. (5.34) is used for calculation of the probability of the sewer system failure ($P_f(t)$), considering all four possible failure modes.

5.4.4.1 Results and Analysis

The results of using first passage probability method (Section 4.2) for calculation of the probability of system failure is shown in Fig. 5.34.

The developed algorithm for GDD model in Section 4.3.3 is used for calculation of the probability of system failure. The results of using gamma distributed degradation (GDD) model for calculation of the probability of the sewer system failure are shown in Fig. 5.35.

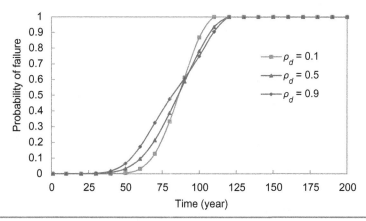

Figure 5.34 Probability of system failure from first passage probability method.

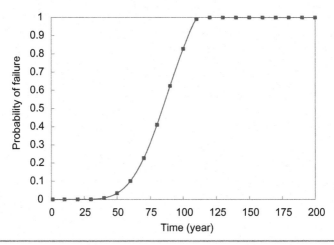

Figure 5.35 Probability of system failure from GDD model.

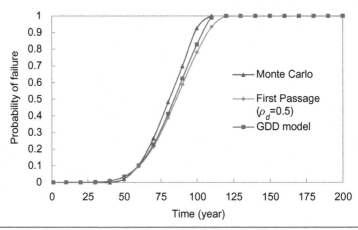

Figure 5.36 Verification of the results from the two methods by Monte Carlo simulation method.

The Monte Carlo simulation method (see Section 4.4 for the detail) is performed for verification of the results from the two methods in the previous section. Fig. 5.36 shows the comparison of the results of the probability of system failure from the three methods (first passage probability, GDD model, and Monte Carlo simulation). The graph for first passage probability method is taken from the result for autocorrelations equal to 0.5.

The comparison shows that the probabilities of system failure predicted by the two methods are in good agreement and it can be verified by the results of the Monte Carlo simulation method, particularly for small probabilities which are of most practical interest.

5.4.4.2 Sensitivity Analysis

As it was mentioned earlier, for a comprehensive pipeline assessment, the effect of variables on the failure of the concrete sewer can be analyzed by performing sensitivity analysis. In view of the large number of variables that affect the corrosion process, and hence the limit state functions, it is of interest to identify those variables that affect the failure most so that more research can focus on those variables.

Unlike individual failure mode assessment (Section 5.4.1), the concept of relative contribution can not be used for multifailure mode analysis. It is simply because Eq. (3.9) has been presented for calculation of relative contribution in case of individual limit state function. Hence, a new parametric method is developed and applied for sensitivity analysis of the concrete sewer in case of multifailure mode assessment.

To assess how the change in the values of the six random variables (k, u, j, [DS], $b/\acute{P}$, and A) can affect the service life of the concrete sewer system, the values for each variable are changed from $\mu_i - 2\sigma_i$ to $\mu_i + 2\sigma_i$ (where μ_i is the mean of the random variable and σ_i is its standard deviation). Assuming a Gaussian distribution for the random variables, this range corresponds to 95.4 percent of the possible values of the variable. The results of the analysis by using Monte Carlo simulation method are illustrated in Fig. 5.37(a–f).

It can be concluded that among all variables, the effect of [DS], A, and $b/\acute{P}$ on the probability of failure of the sewer is highly remarkable. The disparity shown within the graphs for each of these three variables means that the sensitivity of the failure of the pipeline is more dependent on their actual value. In such cases, more concern should be taken in order to determine relevant parameter values.

For a better comparison of the effectiveness of random variables on the service life of the pipeline, the results in are summarized in the form of Fig. 5.38. In this figure, a range value is defined for each random variable as the difference between the maximum and minimum values of the probability of failure in each elapsed time. Therefore a higher range value means wider resultant values for the probability of failures. This figure illustrates which variables contribute most to the probability of the failure of the system. To clarify, this means that the variables have more effect on the service life of the pipeline.

The significance of the three major variables (i.e., [DS], $b/\acute{P}$, and A) on the failure of the concrete sewer has also been concluded in individual failure mode analysis in Section 5.4.1.2.

Of these three major random variables, the analysis shows the particular significance of dissolved sulfide concentration in the probability of failure of the pipeline, indicating that dissolved sulfide concentration is the most significant variable on the reliability of concrete sewers.

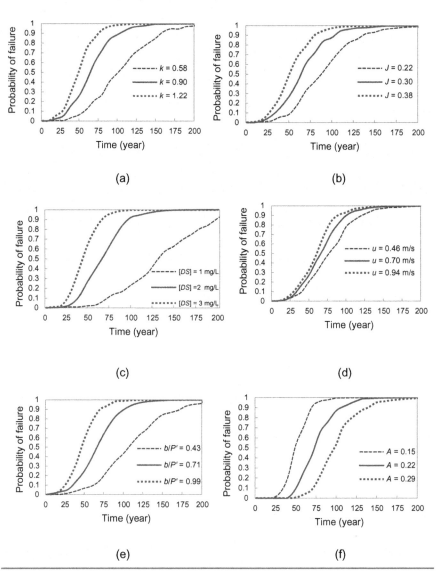

Figure 5.37 Variations in system failure probability of the concrete sewer due to change in (a) acid reaction factor; (b) pH-dependent factor; (c) dissolved sulfide concentration; (d) velocity of the stream; (e) the ratio of the width of the stream surface to the perimeter of the exposed wall; (f) acid consuming capacity of the wall material.

The results from the two developed methods in Chapter 4 (i.e., first passage probability method and GDD model) showed a significant agreement with the results from the Monte Carlo simulation method. Although the result from first passage probability method depends on the amount of assumed autocorrelation coefficient (ρ), in this case study the effect of autocorrelation coefficient on the probability of failure was negligible.

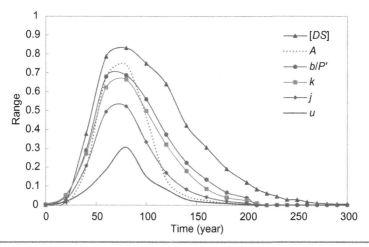

Figure 5.38 Comparison of range values of basic random variables.

In addition to the assessment of existing pipes, the reliability analysis methods used in this case study can also show how changes in design parameters of sewers (such as concrete cover) can affect the service life of the pipeline.

Sensitivity analysis was undertaken to identify factors that affect the probability of pipe failure due to corrosion. The analysis showed less significant contribution of some variables in failure functions. Therefore, it would not be necessary to consider those parameters as random variables and they can be treated as deterministic constant values for further studies.

The results showed that among six random variables, $[DS]$, $b/\acute{P}$, and A have the most effect on the probability of sewer failure. This effect is more considerable for lower values of time, which means special attention should be taken for accurate determination of these variables for new pipelines.

The methods in Chapter 4 were applied to the reliability analysis of a UK concrete sewer case study. The results were verified by using a Monte Carlo simulation, and more realistic results from multifailure assessment were achieved. The degree of sensitivity to different variables was also investigated.

References

Ahammed, M., Melchers, R.E., 1994. Reliability of underground pipelines subject to corrosion. J. Transport. Eng. 120 (6), Nov/Dec.

Ahammed, M., Melchers, R.E., 1997. Probabilistic Analysis of Underground Pipelines Subject to Combined Stress and Corrosion. Eng. Struct. 19 (12), 988–994.

American Water Works Association. (2005). "Installation of Ductile-Iron Water Mains and Their Appurtenances." AWWA C600, Denver.

Anon. 2002. Office of Pipeline Safety, Available from: www.ntsb.gov and www.ops.dot.gov.

ASCE 15-98, 2000. "Standard Practice for Direct Design of Buried Precast Concrete Pipe Using Standard Installations (SIDD)", Reston, VA.

ASCE Manuals and Reports of Engineering Practice - No. 69, 1989. Sulphide in Wastewater Collection and Treatment Systems. American Society of Civil Engineers.

ASCE Manuals and Reports of Engineering Practice No. 60, 2007. Gravity Sanitary Sewers, 2nd edition American Society of Civil Engineers, New York, USA.

BS78-2, 1965. Specification for Cast Iron Spigot and Socket Pipes (vertically cast) and Spigot and Socket Fittings. British Standards Institution.

BS 9295, 2010. Guide to the Structural Design of Buried Pipelines. British Standards Institution, London.

Clair St., A.M., Sinha, S., 2012. State-of-the-technology review on water pipe condition, deterioration and failure rate prediction models. Urb. Water J. 9 (2), 85−112.

Cosham, A. and Hopkins, P., 2004. The assessment f corrosion in pipelines − guidance in the pipeline defect assessment manual (PDAM), Pipeline pigging and integrity management conference, 17-18 May, Amsterdam, The Netherlands.

EPB 276, 2004. Water Pipeline Design Guidelines. Water Security Agency, Regina, Saskatchewan, Canada.

Gabriel, L.H., 2011. Corrugated Polyethylene Pipe Design Manual and Installation Guide. Plastic Pipe Institute, USA.

Gorenc, B.E., Syam, A., Tinyou, R., 2005. Steel Designers' Handbook. University of New South Wales Press (UNSW Press), Australia.

Kiefner, J.F., Vieth, P.H., 1990. Evaluating Pipe: New Method Corrects Criterion for Evaluating Corroded Pipe. Oil and Gas Journal August 6, 1990.

Kołowrocki, K., 2008. Reliability of Large Systems. John Wiley & Sons, Ltd.

Kucera, V., Mattsson, E., 1987. Atmospheric Corrosion. In: Mansfeld, F. (Ed.), Corrosion Mechanics. Marcel Dekker Inc, New York.

Laham, S. Al, 1999. Stress Intensity Factor and Limit Load Handbook. Structural Integrity Branch British Energy Generation Ltd, EPD/GEN/REP/0316/98, ISSUE 2.

Leygraf, C., Wallinder, I.I., Tidblad, J., Graedel, T.E., 2016. Atmospheric Corrosion. John Wiley & Sons, New York.

Li, C.Q., Mahmoodian, M., 2013. Risk Based Service Life Prediction of Underground Cast Iron Pipes Subjected to Corrosion. Journal of Reliability Engineering & System Safety 119, 102−108. November.

Li, C.Q., Melchers, R.E., 2005. Time-dependent reliability analysis of corrosion-induced concrete cracking. ACI Struct. J. 102 (4), 543−549.

Mahmoodian, M., 2013. Reliability Analysis and Service Life Prediction of Pipelines. University of Greenwich, London [Ph.D. thesis].

Mahmoodian, M., Alani, A., 2014. A gamma distributed degradation rate (GDDR) model for time dependent structural reliability analysis of concrete pipes subject to sulphide corrosion. Int. J. Reliab. Saf. 8 (1), 19−32.

Mahmoodian, M., Alani, A., 2013. Multi failure mode assessment of buried concrete pipes subjected to time dependent deterioration using system reliability analysis. J. Failure Anal. Prevent. 13 (5), 634−642.

Mahmoodian, M., Alani, A.M., 2015. Time-Dependent Reliability Analysis of Corrosion Affected Structures. Numerical Methods for Reliability and Safety Assessment. Springer International Publishing.

Mahmoodian, M., Li, C.Q., 2016a. Stochastic failure analysis of defected oil and gas pipelines A2 - Makhlouf, Abdel Salam Hamdy. In: Aliofkhazraei, M. (Ed.), Handbook of Materials Failure Analysis with Case Studies from the Oil and Gas Industry. Butterworth-Heinemann, Chapter 11 -.

Mahmoodian, M., Li, C.Q., 2016b. Structural integrity of corrosion-affected cast iron water pipes using a reliability-based stochastic analysis method. Struct. Infra. Eng. 12 (10).

Mahmoodian, M., Li, C.Q., 2017. Failure assessment and safe life prediction of corroded oil and gas pipelines. J. Pet. Sci. Eng 151, 434−438.

Mahmoodian, M., Li, C.Q., 2018. Reliability based service life prediction of corrosion affected cast iron pipes considering multi failure modes. ASCE J. Infrastruct. Syst 24 (2), 04018004.

Marshall, P., 2001. The Residual Structural Properties of Cast Iron Pipes - Structural and Design Criteria for Linings for Water Mains. UK Water Industry Research.

Melchers, R.E., 1999. Structural Reliability Analysis and Prediction, 2nd Edition John Wiley and Sons, Chichester.

Melchers, R.E., 2009. Experiments, Science and Intuition in the Development of Models for the Corrosion of Steel Infrastructure, NOVA. The University of Newcastle's Digital Repository.

Moglia, M., Davis, P., Burn, S., 2008. Strong exploration of a cast iron pipe failure model. Rel. Eng. Sys. Safe. 93, 885−896.

Moser, A.P., 2010. Buried Pipe Design. McGraw-Hill Professional Pub.

Moser, A.P., Folkman, S., 2008. Buried Pipe Design. McGraw-Hill Professional Pub.

Nessim, M.A., Pandey, M.D., 1996. Risk-based Planning of inspection and maintenance of pipeline integrity. Proceeding of 9th Symposium of Pipeline Research. Pipeline Research International, Houston, TX, pp. 10-1−10-18.

Papoulis, A., Pillai, S.U., 2002. Probability, Random Variables, and Stochastic Processes, Fourth edition McGraw-Hill, New York, p. 852.

Rajani, B., Makar, J., McDonald, S., Zhan, C., Kuraoka, S., Jen, C.K., et al., 2000. Investigation of Grey Cast Iron Water Mains to Develop a Methodology for Estimating Service life. American Water Works Association Research Foundation, Denver, Colo.

Sadiq, R., Rajani, B., Kleiner, Y., 2004. Probabilistic risk analysis of corrosion associated failures in cast iron water mains. Reliab. Eng. Syst. Saf 86 (1), 1−10.

Sheikh, A.K., Hansen, D.A., 1996. Statistical modelling of pitting corrosion and pipeline reliability. Corr. Sci. 46 (3), 190−197.

Sinha, S.K., Pandey, M.D., 2002. Probabilistic Neural Network for Reliability Assessment of Oil and Gas Pipelines. Comp.-Aid. Civ. Infra. Eng. 17, 320−329.

Thacker, B.H., Light, G.M., Dante, J.F., Trillo, E., Song, F., Popelar, C.F., et al., (2010), Corrosion Control in Oil and Gas Pipelines, Southwest Research Institute, San Antonio, TX, Vol. 237 No. 3.

The World Factbook 2010. Central Intelligence Agency, USA, https://www.cia.gov/library/publications/the-world-factbook/fields/2117.html

Yamini, H., 2009. Probability of failure analysis and condition assessment of cast iron pipes due to internal and external corrosion in water distribution systems, PhD dissertation. University of British Colombia.

Further Reading

ASTM E632-82(1996) Standard Practice for Developing Accelerated Tests to Aid Prediction of the Service Life of Building Components and Materials.

Maes, M., Dann, M., Salama, M.M., 2008. Influence of grade on the reliability of corroding pipelines. Reliab. Eng. Sys. Saf 93, 447−455.

Pandey, M.D., 1998. Probabilistic models for condition assessment of oil and gas pipelines. Int. J. Non-dest. Test. Eval. 31 (5), 349−358.

Qin, H., 2014. Probabilistic Modeling and Bayesian Inference of Metal-Loss Corrosion with Application in Reliability Analysis for Energy Pipelines, PhD dissertation. The University of Western Ontario.

Teixeira, A.P., Soares, C.G., Netto, T.A., Estefen, S.F., 2008. Reliability of pipelines with corrosion defects. Int. J. Press. Vessels and Piping 85 (2008), 228–237.

WSAA, Water Services Association of Australia, 2009. National Performance Report.

Index

Printed in the United States
By Bookmasters